SPORTS SPONSORSHIP AND BRAND DEVELOPMENT

Sports Sponsorship and Brand Development

The Subaru and Jaguar Stories

Martin Beck-Burridge and
Jeremy Walton

palgrave

First published 2001 by
PALGRAVE
Houndmills, Basingstoke, Hampshire RG21 6XS and
175 Fifth Avenue, New York, N.Y. 10010
Companies and representatives throughout the world

PALGRAVE is the new global academic imprint of
St. Martin's Press LLC Scholarly and Reference Division and
Palgrave Publishers Ltd (formerly Macmillan Press Ltd).

ISBN 0–333–92540–8

This book is printed on paper suitable for recycling and made from fully managed and sustained forest sources.

A catalogue record for this book is available from the British Library.

Library of Congress Cataloging-in-Publication Data

Beck-Burridge, Martin.
    Sports sponsorship and brand development : the Subaru and Jaguar stories / Martin Beck Burridge & Jeremy Walton.
        p. cm
    Includes bibliographical references and index.
    ISBN 0-333-92540-8 (cloth)
    1. Sports sponsorship. 2. Automobile racing—Economic aspects. 3. Automobiles—Marketing. 4. Fuji Jåkågyå Kabushiki Kaisha. 5. Jaguar Cars Ltd. I. Walton, Jeremy, 1946- II. Title.

GV716 .B43 2001
338.4'379672—dc21

                                                            2001046007

10 9 8 7 6 5 4 3 2 1
09 08 07 06 05 04 03 02 01 00

Formatted by
The Ascenders Partnership, Basingstoke

Printed and bound in Great Britain by
Creative Print and Design (Wales),
Ebbw Vale

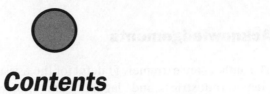

Contents

Acknowledgements

The authors are extremely grateful for the assistance and co-operation of Fuji Heavy Industries, and Jaguar Cars Ltd. Specifically Subaru Tecnica International and Prodrive gave generously of their time and practical help. As did key personnel at Jaguar Daimler Heritage Trust and TWR Group. Without them this case study would not have been completed.

At Fuji Heavy Industries, special thanks are due to Mr Katsuhiro (Keith) Yokoyama, Manager of Corporate Planning Department and Secretary to the President of Fuji Heavy Industries for the excellent arrangements he made and for his patience and forbearance with Martin Burridge.

Jeremy Walton would like to thank the following individuals for their time and co-operation: Jaguar Cars Sales & Marketing Director, Roger Putnam, and Jaguar Daimler Heritage Trust (JDHT) General Manager, Howard Davies. Also the following staff from Jaguar Communications and Public Affairs; Stuart Dyble for his enthusiastic support, Martin Broomer for his guidance, Cecile Simon for his practical assistance beyond the call of duty and Sally Lane, Manager Corporate Affairs. At TWR, our particular thanks go to Paul Davis who supplied production statistics that the authors had not seen previously.

Martin Beck-Burridge is especially grateful to the following executives of Fuji Heavy Industries and Subaru Tecnica International who were interviewed during December 1999. Mr Takeshi Tanaka, President of Fuji Heavy Industries, Mr Takemasa Yamada, President of Subaru Tecnica International, Mr Noriyuki Koseki of Fuji Heavy Industries and Subaru. Mr Kuze and Mr Kazuyuki Narita were kind enough to meet the author in Tokyo and relate their experiences from the start of the Subaru and Prodrive relationship. Also, particular thanks to Mr Keith Yokoyama for his patience and efforts over many months.

Many others helped with information and their experiences and the authors are grateful to them all for their frankness and co-operation.

Any errors or omissions are the responsibility of the authors.

*To Thomas
and our grandchildren*

PART I
THE BACKGROUND

1

Introduction

THE world's automobile industry has, for at least the last two decades, been characterized by over-capacity of production in all sectors, an excess supply of products to all market segments and intense competition within the industry. Manufacturers have been constantly pursuing strategies and policies that will provide them with opportunities for escaping the dominance of price competition, which is still limiting the profit margins of even the largest mass manufacturers. However, for lower volume manufacturers the problem is compounded, because they are unable to pursue the same economies of scale production and cannot compete against the major mass manufacturers on price alone. They are forced to pursue more selective and aggressive policies of differentiation and originality in order to succeed, and therefore avoid facing the mass manufacturers head-on.

The large-scale mass manufacturers and the smaller niche players in the automotive sector compete in global markets characterized by factors that arc redolent of the industry's one hundred years of history, particularly in the major markets of the United States of America and Western Europe. These can be summarized as follows:

- Mature markets with massive over-capacity of supply, probably 25 per cent in 2000
- Low demand growth, 35.1 millions in 1999 to 39.4 millions in 2000; capacity growing faster than demand
- Mass manufacturers increasingly searching for specialist niches for differentiated products
- Product and production technology leading to commoditization of the product
- Growing wealth, information and knowledge of customers leading to diverse demands for individual products
- Intense and increasing market fragmentation

These trends have led to increasing rivalry between the mass manufacturers themselves as they have struggled to increase market share in already saturated markets. In addition the smaller manufacturers such as Subaru, Porsche and others, have focused on being niche players, as the larger companies utilize modern production technology to deliver increasing quality, diversity of product and speed of response. This increases the competition for the smaller niche manufacturer, who has to pursue new means of differentiation and avenues for profitable manufacture, as the larger players probe the smaller niches. The difficulty for the smaller manufacturers is that they lack the opportunity to use large-scale mass markets to reduce unit costs and increase purchasing efficiency, and therefore cannot compete on price. They have to drive costs down through lean production, common components and standardization. Product differentiation and improving the value equation are essential prerequisites to profitability – and the image of the brand is the key.

But the problem is that niche market products can, in some circumstances, be replicated, as advances in product and design technology speeds the process of product innovations. The keys to success, particularly for the low volume manufacturer, are to move from high volume, low value added models to products that, in the words of Gary Hamel, Professor of Strategic Management at London Business School:

- Improve the value equation
- Separate form from function
- Achieve joy of use

The question 'How?' has been answered in different ways. Companies such as BMW have assumed the premium position in each of the sectors in which they compete and pursued a very consistent marketing policy that has given them a clear brand distinction; a marque of aspiration. Within thirty years the company transformed itself from a minor, obscure branded manufacturer into a major worldwide brand renowned for its design, technical excellence and commitment to driving pleasure. Another approach is that adopted by a smaller producer Porsche, who have concentrated on building exclusive sports cars, whilst also engaging in a large volume of high value research and development work for other manufacturers. This policy has, so far, enabled them to retain their profitability and independence, in spite of several unsuccessful forays into larger volumes.

Other manufacturers, notably Ford, Fiat, General Motors and Volkswagen, have vigorously pursued the acquisition and development of valuable niche car brands such as Bentley, Rolls Royce, Aston Martin, Ferrari, Lamborghini, Jaguar, SAAB and Volvo, to name but a few. The point of this is that it seems that we are at the end of a chapter, as the up-market marques are not able to survive as independents, but the up-market brand will survive as part of a mass manufacturer.

Smaller players cannot easily replicate these approaches to intense market competition as they lack the capital resources All of the companies in the previous examples have established strong brands that underpin and indeed strengthen the stance they take in the market place. The combination of exclusivity, product development, improving the consumers' (potential and actual) product perceptions and strong branding is an approach that has led to the realization that 'brands' have actual financial value, as well as consumer value. The creation of brands has become a new phenomenon and companies, particularly those that are fighting to distinguish themselves from the intense competition in industries characterized by oversupply, are determined to imbue their brand with a unique, recognizable and distinct set of characteristics, the 'values of the brand'. But how?

The creation of a brand is, like many management problems, easy to analyse if successful, but impossible to recreate in any laboratory. Laboratories for the *post hoc* analysis of management strategies and methods do not exist; the replication of concepts and the repeated experimentation available to the physical sciences are simply not available. So, whilst the examination of successful and unsuccessful strategies is valuable, the answers provide no guarantee of success. However, achieving success in brand definition is a well tried path to elevating a company's products into market spaces where premium prices can be obtained and products differentiated in an overcrowded market.

An effective and dominant brand has the added advantage to the producer of 'reaching over the shoulder of the retailer straight to the consumer' – a phrase coined by H G Wells – and therefore is part of the process of differentiation, identifying the product from others that may perform a similar function. Branding is an essential part of the differentiation of products in a highly competitive market, products that would be, or are, otherwise regarded as commodities or at least those that are subject to intense price competition. Developing a new brand, or alternatively re-creating an existing brand all involve the identification of distinct characteristics/values that the consumer will recognize and value and yet is different from the characteristics offered by other brands. Moreover, it is preferable if the characteristics are difficult to imitate and therefore provide the consumer with a unique set of values, images and consumer positioning beyond the functionality of the product.

However, in a product-driven industry a brand does not exist in isolation from the product. The creation/recreation of a brand has to be driven by the product's impact upon the customer's perception; the product's definition. Having said that, there are exceptions such as the perfumery industry, in which there are examples of products that rewrite the rules of product definition and brand image. The perfume 'Charlie' is one such example, since perfumes had, until the launch of that fragrance, always been given somewhat romantic and intensely feminine names such as 'Miss Dior' and 'Chanel'. Whereas the perfume 'Charlie' was the embodiment of a rather more assertive, independent and positive female image, that immediately captured the attention and the money of a particular section of an affluent (and younger) consumer market.

The wide range of automobile models that are available to the consumer come at all levels of price, consumer perception, design and performance. It is the brand of the company that creates the initial product perception and the individual products' position in the market place. A BMW 5 Series saloon has a distinct and positive image very different from Ford's Mondeo, or other models of a similar marque and market position. The brand sets the pace, but the product has to be able to fulfil and even extend the brand image and attributes, the benefits and essence that they convey in perception, in their use to the consumer. It is these that are the values of the brand and it is the value in use that the consumer demands. However, that value in use also often means the value that others perceive in the consumer using the article.

Many of the 'brands' in the automobile industry have been created from a position of market weakness, or have survived periods of decline, including Jaguar, BMW and Porsche. BMW had to thrive in the aftermath of the Second World War, Porsche was actually launched at that time and created its image, but the brand has subsequently survived two major recessions in its major markets Europe and the USA. The Jaguar brand has had to survive the self-inflicted destruction of build quality and the 'essence' of its brand, but also the (probably not unconnected) bleak decades of the decline of Britain's motor industry. These companies successfully retained or, in the case of Jaguar, recreated their brand's image. However, in the post-war period the Japanese were building a new industry and therefore Japanese automobile manufacturers had to create an effective brand image rapidly, if they were to compete in Europe and North America.

Unlike the Japanese manufacturers, American and European automobile brands have a long history, in most cases going back to the beginning of the motor car's century. By the end of that century, amongst the top ten brands in the automotive and oil sector, there are two Japanese manufacturers,Ñ Toyota (fourth) and Honda (eighth), three German, one American and one Swedish brand (The World's Greatest Brands, ISBN 0–333–66419–1). As the European and American manufacturers have a forty year start on the Japanese, this and the rise of their global industry is an exceptional story. However, whilst Ford is the only mass manufacturer in the top ten from outside Japan, two out of the four truly global mass manufacturers from Japan are in the top ten; Toyota and Honda, both selling approximately 2.5

million vehicles annually worldwide. That is a remarkable achievement and does perhaps tell us something about the future of the industry.

The creation of a new brand is not necessarily a forty-year project, nor is it always the largest players that succeed. The process of achieving a successful brand within any market is fraught with difficulty and it is said that out of every twenty attempts at launching a new brand or recreating an existing one, seventeen fail. As with so many management issues the problem seems to be that it is perfectly logical to establish reasons for failure, and even the criteria for success in specific circumstances, but to identify necessary and sufficient reasons for success seems to elude even the most insightful marketing analysts. So in certain circumstances how are brands created, what are the necessary conditions for success and indeed, what is a 'brand'?

Brand Development, Sponsorship and Marketing

The rapid and unprecedented economic growth of the twentieth century (*see* National Bureau of Economic Research Working Papers 7569, 7602 by B. DeLong, World Economic Outlook 2001, published by the International Monetary Fund; 'A Century of Progress', *Economist* 15 April 2000) has resulted in equally massive changes in the marketing of goods and services. The concept of consumer brands was born in those decades and the world's consumers now have a more varied and wider set of branded products than at any time in the history of consumerism. The concept of a brand has been transformed from a means through which to differentiate a product or service, with a derived value, to being a major source of value in themselves. There are a number of companies that have developed their businesses through purchasing portfolios of brands, as others have by purchasing products and technologies.

The success of brands, and the fact that they now have a value of their own and can be applied to ranges of products, increasing the differentiation of the product and thereby the value perceived, also elevates the price paid by the consumer. There are a number of companies that have successfully and aggressively built shareholder value on the basis of brand acquisition. This is not necessarily done on the principle of purchasing brands in a specific sector such as automobiles, but often by the criterion of type of customer. The first brand pursuit involved car makers such as Ford, whose pursuit of brands in the late 1990s included the acquisition of Aston Martin, Jaguar and Cosworth, then General Motors' acquisition of a proportion of Subaru and Mazda and the aggressive and relentless path of acquisition pursued by Volkswagen.

Alternatively, one entrepreneur has grown a tired and loss making textiles group that owned the Christian Dior brand into a profitable £28 bn empire known as LMVH. The management of the company has created a profitable and global empire out of the desire for the consuming public to own distinct, differentiated, designer emblazoned articles. The articles say as much about the owner's life and style as about the size of their bank balances; they add to the identity of the user. Bernard Arnault, the architect of the company has seemingly proved to be the 'architect of luxury' who has, until recently, proved to be infallible in his pursuit of style and profit. However, the House of Gucci, riven in the past decade by family feud, scandal and even murder, was purchased last year by his rival style entrepreneur Francois Pinault's company Pinault-Printemps-Redoute (PPR). Both companies are a development of the brand revolution that is one of the new elements of the commercial and financial basis of modern companies.

Both companies have pursued 'style' brands and are less concerned with the type of product than with the style, essence and status of the brands themselves. The image that the brands give the consumer is not simply one of 'expense' but also the other essences of a brand that provide status in terms of taste, style and exclusivity. In addition these brands have made it possible for the companies concerned to build product portfolios around the brands that provide distinctive qualities, premium prices and therefore abnormal profits. However, as has happened so often in pursuing the latter, the brand itself can be too widely applied to unsuitable products that have no 'essence', therefore diluting the brand and the essence, which disappears along with the profits.

One definition of a brand has been stated as follows:

'A mixture of tangible and intangible attitudes, symbolised in a trademark which, if properly managed, creates influence and generates value.'

(*The World's Greatest Brands*)

There is little reason to alter this definition, but the modern, consumer driven world has determined that the word 'brand' is applied to a wider and more eclectic group of products, services and even institutions such as teams and corporations rather than just products. Currently, brands are associated with public (and private) utilities, football and other types of sports teams, charities, product and service-based corporations and, even individuals, as well as products. All of these categories of organization regularly use brands to define themselves and/or their products; they create the brand as the focus of their values and qualities.

The common aspects of all these concepts called brands is the specific characteristics that apply to the most successful brands, characteristics that when isolated and measured can be used to define the character and impact of the brand. Furthermore, when defined these characteristics are also a valuable insight into the reason for the success of particular brands and the key factors in building a successful brand. These elements are (as quoted from *The World's Greatest Brands*:

- CLARITY – of vision, mission and values, which are understood, lived and even loved by the people who deliver them. Clarity of what makes those values distinctive and relevant; and clarity of their ownership in both people's minds and in trademark law around the world.

- CONSISTENCY – not in the sense of a simple product, nor in the sense of predictability. Successful brands are consistent in the values, concepts and level of quality they or their products deliver to the consumer. A consistent alignment of values such as those expressed by Moet & Chandon, Coca Cola, Ford, BMW or Gillette, to give but a few examples.

- LEADERSHIP – a consistent characteristic of successful brands is the brand's ability to lead and exceed expectations, to take people into new territories and new areas of product, service and even social philosophy at the right time. It is about a brand's ability to be restless about self-renewal.

This means that a successful brand is defined firstly, by the clarity with which the organization's values are portrayed by the brand's image and message; secondly, by the consistency that the brand's image is perceived; thirdly, by the brand's ability to redefine the values and meaning of its service or product characteristics and properties. This latter point suggest that brands are evolutionary and indeed need to be able to recreate themselves as social, product, market or indeed any other critical values and conditions change. In this context a brand has to be able to provide a consistent set of values and characteristics as well as an adaptable image that on occasion leads societal values.

The fact that many of the world's top brands have been developed by the largest organizations is not of itself surprising, what is surprising is that many smaller companies have been equally as successful. In an Interbrand survey of the world's top brands by sector (*The World's Greatest Brands*, quoted above) the inclusion or exclusion of brand names is not a direct function of the size of the organization, either by total revenue or any other measure of size. The crucial characteristic is the ability of the product or service, and ultimately the brand, to characterize and communicate specific, probably unique, qualities and values to consumers within the market segment and to a wider audience.

This aspect of branding, the smaller company attempting to build a brand within the wider context of one of the world's largest industries, presents particular problems to a company that is a small player with a national brand, unknown or at least little known within the world's market place. In many industries the successful organizations and their brands have been established by the careful and creative application of the critical principles of brand creation within very clearly defined market niches. Examples of this successful application of consistent, defined values and precise brand characteristics in niche markets are BMW, Harley-Davidson, Rolex, Porsche, Volvo, Financial Times and Benetton all considered to be in the world's top one hundred brands. None of these companies built their brands within large corporate entities. BMW, Harley Davidson and Porsche, who are all still independent, built their brands by focusing on a niche market and concentrating upon the characteristics and values of their customers. In the cases of Volvo and the Financial Times, their success created a brand that had a financial value to the large corporations that purchased them; they were purchased for the strength of their brand image.

The issue for the niche automobile manufacturer is 'how'? How to achieve the development of products that provide the consumer with a product, that is as much a way of life and an expression of shared values as a functional object; functionality is now too easily achievable to be a sustainable profitability proposition.

So how does an automotive company with a small market and a limited brand image, recreate itself and its brand to achieve distinctive and sustainable profitability? The key is in the satisfaction of the differentiated needs and the clear definition of the precise values of the market segment. It is in achieving the key elements of niche marketing, that are: 'Offering distinctive specialist products or services at premium prices, to a population of customers with highly differentiated needs.'

However, the options for providing the consumer with such products are more limited for a niche player in the industry. In view of the huge investment required in vehicle development, the level of marketing and promotion required to bring the products and values of the company before prospective customers is often prohibitive. Moreover development of products is often too long as the keys to success are speed, product differentiation and brand consistency and differentiation; price competition, for a niche player, is death. This leaves the niche player with a number of options to pursue:

• Joint venture with a major manufacturer
• Conceive and build low volume niche vehicles that are produced in-house
• Alliance with a specialist engineering/motorsport Group

The question is which?

2

Sport, Sponsorship and Marketing

The Background

BBC television and subsequently the commercial channels followed a cross-section of British sports in the 1950s and 1960s. As football, tennis, boxing, motorsport and cricket developed a wider TV audience, many of the teams and individuals involved gained increasing amounts of corporate sponsorship, that were not always beneficial. Debate resounded around the offices of sporting organizations, lamenting the loss of the 'True Amateur' and the spirit of the 'game'. Commercialism was not simply influencing the way in which sports were presented, it was altering the financial and organizational structures, both of the participants' involvement and the administration. Now, everyone had to deal with 'The Sponsors'.

Television has an insatiable appetite for any colourful spectacle, but there has to be an alert organization behind the action that is displayed on the small screen. The Formula 1 circus that is now a highly organized spectacle, has at its heart the cluster of engineering excellence that thrives off motor racing and is fuelled by sponsorship monies. The fact that it is located in the United Kingdom is, at least partially, due to historical accident as we explained in 'Britain's Winning Formula'; neither the global show nor engineering business of motorsport could exist without the large and continuous flow of sponsorship funds. Meanwhile, the media concentrates on the personality cults that dominate mass-market sports, building the total Formula 1 show into a worldwide spectacle that attracts an audience far wider and more numerous than purists.

When the commercial transformation of motorsport began in the 1970s, Formula 1 was not the premier category, neither was it the most well organized or popular motorsport division, for the Americans had exploited both sponsorship and multi-media coverage both before and after the Second World War. During the 1950s and 1960s, the F1 team owners were independent, entrepreneurial characters, 'racers' with apparently little interest in the wider organization of the sport, except as it affected their racing. But the leisure and commercial world was changing and becoming global, linked by television and a modern business culture. It became very obvious to at least two people involved in the sport, Bernie Ecclestone and

Max Mosley that this sport and that of the World Rally Championship (WRC) needed refreshed organization and personnel to take it forward. Motorsports would become prime beneficiaries of television's benevolent(?) hunger for spectacle to feed the largest audiences in history.

The Beginnings are Established

During the immediate post-war decades both types of motorsport, rallying and Formula 1, were casually organized, haphazard activities confined to hardy enthusiasts and even more hardy practitioners. Advertising at each circuit was the responsibility of the circuit owners and the main sponsors at both types of event were almost always automotive suppliers or car dealers. The circuit owners would be uncertain as to which teams and which cars or drivers would actually appear, let alone race. The money involved was relatively small as were the amounts earned by the leading post-war drivers and their teams.

Personalities such as Tony Vandervell, Raymond Mays, Peter Berthon and many others contributed massively to the development of the pool of vigorous engineering and driver excellence, building a foundation for the UK motorsport industry (see 'Britain's Winning Formula'). Contributing companies had very little stimulus, or opportunity, to capitalise on their investment. Corporate marketing was an arcane mystery and commercial television was absent until the mid-1950s, and therefore so were the massive global audiences.

By the 1960s, the dominance of the pre-war manufacturers' marques such as Mercedes Benz, Alfa Romeo, and Maserati had waned. Cooper ceased trading in 1968, but had already overturned the front-engine establishment and shown that commercial racing car construction could be a viable business. Lola, Brabham, Lotus, McLaren and then March, all contributed to the demise of the establishment, but even in the post-war seasons, the presence of Ferrari has remained central to Grand Prix's televisual appeal.

Successful, mostly British based, chassis fabricators turned the fabrication of racing cars into a 'kit car' business, at least according to one Enzo Ferrari. However, two fundamental elements that made this possible were the outstanding Ford Cosworth DFV engine, the brainchild of one Walter Hayes and the Cosworth Company, and the affordable and remarkable, Hewland racing gearbox. Colin Chapman originally persuaded Ford to invest in the Grand Prix engine and Walter Hayes had the foresight to commit Ford's money. The Ford Motor Company now own Cosworth Racing, the company that built the remarkable Cosworth Ford DFV engine which, after the 1968 season, was sold to other GP teams.

At the time, the opportunity to purchase potentially winning hardware was vital for the changes to transform motorsport into a business since manufacturers charged huge amounts for engines, if they made them at all. The other critical ingredient was sponsorship, yet another area where Colin Chapman's genius was evident. The John Player Tobacco Company sponsored the 1968 Lotus 49, and the car was painted in the red and white of their new brand of cigarettes, Gold Leaf; thus was the modern corporate sponsorship of motor racing born.

It was 1970 before the next Grand Prix sponsor, the toiletries and perfumery manufacturer Yardley who sponsored the British Racing Motors team in 1970, became probably the first major non-automotive entrant into the rarefied arena of Formula 1. More non-automotive sponsorship, and the increasing number of safety inspected tracks around the world, were crucial elements that transformed Grand Prix, and enabled the last stage to take place. The first stage had taken more than twenty years from the end of the Second World War, the next phase would take only half as long. In that decade, an activity dominated by enthusiasts would be transformed into a spectacle that would span the world and involve practically all of the business sectors, from computers to oil, cars to clothing, management consultants to newspapers and drinks manufacturers.

The Development of Organized Spectacles

F1 and the WRC people

In fact for a number of historical and economic reasons, Formula 1 Grand Prix racing was to be the formula that led the development of the motorsport business. After the success of the reorganization, some would argue simply organization, of Formula 1, parallel changes based on that example, of providing a reliable show of competitive quality, were adopted by other branches of motorsport. That has been particularly true of the World Rally Championship (WRC), which had no World Champion driver series until 1979 and did not field all the competitive cars and top teams until the later nineties.

Even then, WRC television coverage was sporadic and deeply disappointing to the major manufacturers. However, when in 2000, David Richards the Chairman of Prodrive led an independent financial deal to buy the FIA Television rights to the WRC, it appeared that things would change.

The acquisition was a direct result of a strategic review of Prodrive undertaken in 1998, which had led to American venture capital company Apax acquiring 49 per cent of Prodrive and providing the capital for expansion. The purchase of International Sports World Communications, originally a part of the Ecclestone empire, gave the WRC TV Rights and unspecified 'certain Commercial rights' until 2010, to the group led by Richards. The new entity has a number of commercial partners, including Chrysalis TV, Haymarket Publishing (publishers of the RallyXS magazine) and Sony, the licence holders for WRC computer games who are launching the WRC Playstation2 game in October 2001.

There is little doubt that these changes will provide a commercially exciting opportunity for the World Rally Championship to become a competitor for the worldwide TV audiences that excite sponsors. There is also no doubt that the success of International Sportsworld Communications is dependent upon its ability to attract the audiences and therefore the sponsors, and that will be a function of Richards and his team at ISC achieving their stated objective, which is to: 'Deliver compelling TV content to achieve the position of key global media content provider by 2005'.

However, to achieve that the company will need to expand rapidly, both along the automotive and non-automotive sponsorship/TV lines of mainstream motor racing, and the early expansion of the WRC is crucial to that goal. Nevertheless, the fact is that the path has already been trodden, and there is an enormous pool of knowledge within the motorsport industry and sponsors, that was non-existent when Ecclestone took Formula 1 into the TV age.

It is too early to say if this change in TV rights, which means British viewers turning from major league BBC Grandstand to the lesser terrestrial audience of Channel 4, will be a success. However, David Richards certainly possesses the credentials to 'do a Bernie Ecclestone' TV conjuring trick. The World Rally Championship is a spectacular sport that Richards has conquered at every level, from participant to multiple World Championship winning entrant. The question is whether or not the sponsors can be persuaded to enter the sport in sufficient numbers, and whether the TV audience can be expanded.

Chronicling the roots of the changes in the Grand Prix circus, is a book written by Alan Henry some twelve years ago entitled 'March: the Grand Prix and Indy Cars' (1989) which recounts important aspects of the story of the business expansion of Grand Prix. Max Moseley started March with co-founders Robin Herd, Alan Rees and Graham Coaker. As Max Mosley has said:

Some rounds of the 1969 World Championship had only 13 cars. All this changed in 1970, with ten March Formula One cars built and usually five or six of them on the grid. In addition, everyone could see there was nothing magic about Formula One, if we could start in September with an empty shed and no money yet be on the front row of the grid the following March, so could they.

Previously, everyone believed that either exotic genius or large automotive corporations were prerequisites to Grand Prix success. The current Formula 1 drivers' Championship started in 1950, and that for constructors in 1958. They provided a focus for drivers and teams alike, but there was no rule insisting that participating teams presented cars at every round of the championship, nor were individual events co-ordinated.

In 1969 Bernie Ecclestone offered to help establish a new Formula 1 Team with Jochen Rindt and Robin Herd, each having 45 per cent of the equity. This did not happen, but by 1970 the March Team had entered Formula 1 sponsored by the major oil additive company STP. This gave the urbane, intelligent and extremely shrewd Max Mosley a place on the Formula 1 Constructors Association, FOCA.

Max Mosley is the barrister son of Sir Oswald Mosley and a man of exceptional determination, eloquence and intelligence. The authors had an extended interview with him in April 1998, when we discussed the past and future business of Formula 1 and other motorsports; it was apparent that Mosley's fervour for this sporting business would seem to transcend the commitment of the dedicated businessman or barrister. Mosley emphasized that the modern organization of Formula 1 Grand Prix was not an accident, but the vision of Bernie Ecclestone, which had probably started in 1957. In that year Ecclestone had purchased the assets of Connaught Engineering at auction, only to withdraw from racing a year later after two driver fatalities in 1959. A decade later in 1969 Bernie Ecclestone, then running Colin Chapman's Formula 2 team, was enthusiastic about returning to Formula 1.

That opportunity arrived in 1970 when he purchased an interest in Motor Racing Developments, the Brabham Team, from Ron Tauranac. Subsequently, in October 1971 Ecclestone purchased the Brabham team outright from the designer and part owner Ron Tauranac. After Tauranac's departure Ecclestone reportedly said:

The most important thing for anyone in business is to be able to make decisions quickly ... This means that the top job is a one-man job. If there are two top men a lot of time is wasted just trying to agree. (*Eric Dymock 1980*)

This event marked the beginning of Ecclestone's influence on the future of Formula 1, as he now had a seat in the Formula 1 Constructors Association (FOCA). Change did not take long to begin, but it was to be a turbulent eleven-year haul before the vision was implemented. Throughout the following seven years the Federation Internationale du Sport (FIA) and FOCA engaged in frequent conflicts over practically every aspect of the presentation and organization of the sport. Jean-Marie Balestre was the main spokesman for the FISA, as head of the CSI committee, whilst throughout this period Max Mosley and Bernie Ecclestone represented FOCA on the FISA Formula 1 Commission, negotiating with Balestre and the other members. Balestre wanted to place commercial control of the sport back into the hands and control of FISA, and by the end of 1977 the antipathy between the two parties was clear. In 1978 Balestre was elected President of FISA and the power struggles went public.

The Organization

At the 1972 Monaco Grand Prix, FOCA clashed with the International Sporting Commission (the CSI) over the number of cars on the grid. The increased commercialism of the sporting spectacle and the pressure from sponsors, had multiplied the number of cars involved to twenty-five, whilst the AC de Monaco had mandated a grid total of from 16 to 18 cars. The constructors association decided to make a stand over the seven barred entries, the result of the stand-off being that after prolonged discussion they negotiated the 'Geneva Agreement' that established a set payment to each team, guaranteeing a full grid for each race.

The Constructors were also unhappy about technical regulation changes, but the real issue, money, did not surface until October 1972. The circuit owners were represented by an organization called Grand Prix International, and the Constructors had issued a demand for increased appearance money, raising the appearance fee for each team to £103 000. The GPI and the CSI had a more or less common membership base and they worked together to resist the demands of the Constructors, responding to FOCA with an offer of £53 000.

In the winter of 1972, the root causes of the decade of disputes between the Formula 1 Constructors Association and the GPI became obvious. When the Formula 1 circus arrived at Montjuich Park in Spain for the first of the European races, three members of FOCA's finance committee, Max Mosley, Bernie Ecclestone and Phil Kerr, decided to ignore the GPI and negotiated with individual circuit owners. They were not going to give up the Grand Prix gold that their organization was about to create.

FOCA's organization had taken the initiative by refusing to be drawn into negotiations and the CSI and GPI were on the defensive. As the 1974 season started, the problems created by the oil crisis did not diminish the commercial interest and financial support that the Formula 1 circus received. Sponsorship encouraged more non-automotive sponsors. In spite of all the arguments, and a number of fatal accidents to drivers, Formula 1 continued to thrive.

The next season, 1975, was an important milestone in this third phase of events that changed the face of Formula 1. The battle between the CSI, FOCA and the GPI flared as the Grand Prix Drivers Association (GPDA) lost the safety battle with the team owners at the Spanish Grand Prix, in spite of a drivers' strike. Whilst the drivers began to work under the umbrella of FOCA, CSI proved to be an almost spent force in this dispute. In spite of the death of four people when a barrier failed and a car went over it into the crowd, the racing went on and yet the barriers failed yet again. The Constructors then cancelled the Canadian Grand Prix, yet another example of the inadequacies of the CSI.

By the end of the 1975 season, a deal had been worked out between the CSI and FOCA that was to govern Formula 1 for the next decade. The deal covered a 26 car grid (there are 22 on the grid in 2001) and organizers paid a set rising fee for each racing season, the administrators at FOCA being responsible for allocating the money to individual teams. The CSI set up a special Formula 1 committee including representatives from FOCA to undertake the government of the circus. To quote Eric Dymock in 1980: 'Having subdued the drivers FOCA now proceeded to subdue the CSI.'

A further seven years of uneasy and unsettled progress commenced.

Whilst Bernie Ecclestone and the FOCA organization had taken the initiative from the CSI, Balestre, and therefore FISA, and perhaps more importantly Ecclestone, had clarified the process of staging a Formula 1 race, and made the lives of the team owners easier and more lucrative. He had negotiated a deal with FOCA, where they contracted Ecclestone's organization to set up the shipment of the equipment and cars, the race calendar, and the race track facilities (including on-track hospitality, advertising and all television rights), in return for a fixed fee. Ecclestone's organization became the promoters of Formula 1, acting on behalf of the teams, and Bernie became the sport's commercial director. The actual programming of races on the television, was still a haphazard affair with most companies wanting to be paid to broadcast races.

Talking in April 1998, Max Mosley recalled that by 1977 the situation had reached a pivotal point, when the expense of putting on a Grand Prix was a major issue for the circuit owners, who had become nervous of last minute cancellations. At least one circuit group had offered the Grand Prix directly to FOCA, rather than run risks of the uncertainty. This could have meant that the mainly British based constructors would have been running a separate and therefore rival set of races, which was commercially impractical.

Between 1977–80 the number of Formula 1 team sponsors increased, funding the money the constructors were spending, and the wealth of the sport and the people in it. The sponsors ranged from tobacco companies to automotive product makers, perfume and toiletry manufacturers, toy companies, a bank and a manufacturer of audio-visual equipment. Sponsorship money was not yet magnified by the wealth of global television. There were a few more changes that the management of the sport would have to implement before the organization would deliver.

The struggles between the constructors and the international body that controlled motorsport, the Federation Internationale de L'Automobile (FIA) and its subsidiary body, the Federation Internationale du Sport Automobile (FISA) had become a saga, with Ecclestone and Jean-Marie Balestre as leading characters and protagonists. Naturally, the personalities were as important as the issues involved.

The dispute between British-based 'kit car racers' as they had been dubbed dismissively by Enzo Ferrari, and the major continental manufacturers, Fiat-funded Ferrari and Renault teams, simmered on with arguments about sliding skirts and other regulations. The obvious compromise of FISA dealing with the regulations and FOCA dealing with the money and commercial aspects appeared impossible. Firstly, because poorly worded regulations placed huge amounts of sponsorship money and the ability to provide the racing spectacle, at risk. Secondly, Balestre was determined that the commercial aspects of the sport would be taken back by the FIA.

These matters were certainly at the forefront of FOCA's thinking in relation to the turbocharged engines. Teams without access to turbocharged factory engines could, and probably would, become uncompetitive and therefore place themselves and the circus at financial risk. Formula 1 regulations continue to remain fundamental in the same way as governmental regulations affect industry, only far more so since the sport's competition is essentially driven by the manner in which engineers interpret the regulations; F1 is essentially a free formula.

The 1981 Concorde Agreement

Balestre understood that FISA had disregarded its responsibility to regulate in a manner that ensured their rigorous and unbiased implementation. FISA's attempts to ban ground-effect aerodynamics failed because the ingenuity of the designers, in particular Gordon Murray, was superior to that of the legislators. The television income was an important aspect of the Concorde Agreement of 1981, encouraging the protagonists to settle their disagreements. It was apparent that the sponsors and many of the suppliers did not want to see motor racing becoming like boxing, with a plethora of championships.

Thus representatives of both FISA and FOCA attended a meeting in January 1981, at the Ferrari-dominated heart of Italian motor racing, Maranello, Italy. The Concorde Agreement of 1981 was signed, no doubt to the deep relief of all concerned, effectively splitting the running of the sport into the matter of financial negotiations for the race series, and the regulations governing the technical aspects of the races and racing cars. FISA was the acknowledged authority on all aspects of the technical rules governing the races and the cars, whilst FOCA would be responsible for the financial aspects.

Under the Concorde Agreement of 1981, Balestre effectively leased the TV rights to FOCA as the contract acknowledges that the FIA owns the TV rights to Grand Prix races, forever. The actual contract was with FOCA, not with Ecclestone, who was effectively subcontracted through his own company to provide the commercial management of F1. Ecclestone had created an effective route to solving the complex problem of managing the regulatory and commercial aspects of Grand Prix. Balestre eventually made Bernie Ecclestone a Vice President of the FIA, allowing Ecclestone an ideal access to FIA members of governing committees.

In the ensuing decade, the acrimony between Balestre and Ecclestone surfaced frequently. When the March team pulled out of F1, and eventually out of business, Max Mosley continued to work in motorsport as a freelance consultant. Mosley developed a sound knowledge of the workings of the FIA, and a clear understanding that, in common with many other sports, the FIA was run by a group of talented amateurs, in an era of professionals. Mosley was not alone in thinking he could do a better job.

In common with the other protagonists, the FIA members saw the destructive and unsettling results of the acrimony that Balestre and the Formula 1 Constructors Association created, as did the circuit owners who often bore the brunt of the disagreements. The FIA were, of course, aware that much of their income came from the TV rights and other fees generated by the efforts of Ecclestone and Formula 1. In 1986, with the organized support of Bernie Ecclestone, Max Mosley stood for President against Jean-Marie Balestre. Max was also backed by Formula 1 team owners, plus a large number of the 143 clubs from 117 FIA member countries.

The Englishman surprised no one, except perhaps Balestre, when he won convincingly. Subsequently Max Mosley admitted that his careful assessment of FIA operational relationships did not fill the forty-six year old barrister with confidence. Mosley identified three key issues as crucially bad aspects of the contractual relationships between the FIA and FOCA.

Under the 1981 Concorde Agreement, the FIA had a legally loose and untidy contract with FOCA (Formula 1 Constructors Association) and, as Mosley pointed out, this was a changing group of teams that could not be a stable group. Secondly, the very source of the FIA's income came from Bernard Ecclestone's own organization, and yet the FIA had no contract with that organization at all. Finally, it was the individual teams, that were members of FOCA, who were responsible for attending each of the season's races with two cars, and therefore the full grid depended upon the group's consistency, yet the group changed repeatedly. This was certainly not a situation that was commercially or contractually effective, and yet the whole edifice of the spectacle depended on this collection of probably unenforceable, certainly legally untidy, arrangements: intolerable to this barrister of motorsport.

The mutual respect which Mosley and Ecclestone had, one for the other, was an important factor in the next round of the transformation of Formula 1's business. Mosley recognized the reality that the teams were already paying Ecclestone, who allocated funds from the circuit owners and other sources to the teams. Mosley was convinced that by dealing directly with the man who had developed the vision of Formula 1 Grand Prix, and knew how to ensure that the teams and the circuits delivered consistently, Mosley would be able to ensure the vital income stream for the FIA.

In an article in *Business* in January 1999 Mosley is quoted as saying:

> 'There was a kind of constantly changing bunch of contractual partners, and the man with the real money was hiding behind them. We had to have a contract with Bernie Ecclestone and he was more and more the man with the money, it seemed to me that we needed a contract with Bernie and also with the teams, because of their own actions. If we had a contract with Bernie then it wouldn't matter whether the teams came and went and which teams were there or not.'

Late in 1994 a meeting of the Senate of the FIA, part of the organization that had been established on the recommendation of Ecclestone which had Balestre as Chairman, initiated a far-sighted policy. Over the following two years the situation between the FIA and FOCA was resolved, except for the complication posed by the Competition Laws of the European Union. Ecclestone assumed the contractual responsibility, with a fifteen-year contract for the FIA's TV rights. Whilst the role of FOCA is vital to the business, at the very centre of the business is the FIA Senate and then

Ecclestone's own company, one that was contracted to run the commercial aspects of Formula 1 on a 15 year contract.

Some FOCA members felt that the increasingly massive television revenues that the F1 spectacle generated, were essentially theirs. Max Mosley sticks to the law and the spirit of the original agreement, that all the parties had signed in 1981. He told us: 'The vision is Bernie's and he has generated the TV interests and knows how manage them.'

In addition, and perhaps rather more importantly, there is no one else who knows television, the intricacies of worldwide promotion and the business of motor racing, so thoroughly. The new contract with Bernie Ecclestone's company succeeded the defunct Concorde agreement.

There was another aspect of the FIA's strategy in giving Ecclestone the extended agreement, which was the FIA's desire to see a more structured corporate entity in place that would provide continuous executive management of the commercial side of the sport. The FIA were correct in being concerned with the matter of continuity, as was tellingly confirmed in June 1999.

Bernie Ecclestone was born in 1931 and is over 70 years old. In June 1999 he underwent heart by-pass surgery, from which he has made a full recovery. Nevertheless, given the increasingly rapid growth of the revenues, and the importance of those revenues to the financial health of the FIA, their prudence was logical. But the deal had another twist yet to come.

In 1996, Ecclestone had started to work on the flotation of a bond, based on the increasing revenues earned by SLEC Company, which controls the Ecclestone empire under the family trust, and thereby the revenue stream that was now guaranteed for ten years. These bonds are a relatively new financial instrument, developed to provide people or organizations such as writers, musicians, or inventors with important patents which provide them with a relatively secure and regular income stream, with an opportunity to raise capital sums based upon the security of that income stream.

After prolonged negotiations with several financial institutions, by the end of 1996 Ecclestone had reached an agreement with Salomon Brothers, who had agreed to float the bond for a total of $2 billion, approximately £1.35 billions. This prompted the FIA to offer to increase the period of Ecclestone's television rights by a further 10 years, in exchange for 10 per cent of the bond. Salomon Brothers attempted to float the bond in 1998, the two year gap being due to the internal wrangling over the allocation of the TV revenues to the teams; but the float failed. Further attempts to float the bond were met with considerable market scepticism, and even when Morgan Stanley came alongside, the issue failed to proceed.

Then, in May 1999 the bond was at last given the essential rating by the top four agencies as a prerequisite to issue; this came through the combined efforts of both banks plus the assistance of the WestDeutsche Landesbank. The latter bank persuaded Ecclestone and the FIA to accept that the bond's period be reduced from 20 to 10 years and the issue value reduced by $600 000, bringing the issue price down to $1.4 bn (£880 m). However, the market is still somewhat unsure of the bond's real value, although it has been issued on the Luxembourg Stock Exchange. The European Commission's Competition Directorate announced their preliminary findings on the contracts in June 1999, after a two-year inquiry, as follows:

> The inquiry concludes that Mr Ecclestone broke the EU competition law when he obtained exclusive television broadcasting rights to world motor racing events. European Commission insiders said that the contracts – against which a $1.4 bn (£880 m) Formula 1 bond issue was secured – would have to be renegotiated if the initial findings of a Commission investigation were confirmed.
>
> *Financial Times* 30 June 1999

This problem will probably continue to haunt the bond's future, unless the European Commission's Competition Directorate and the Commissioner Karel van Miert can be convinced to withdraw their objections to the contract, which was probably awarded to Ecclestone's company without any semblance of competitive tendering. The existing contract has been in place for some five years and the Commission's response to the documentation submitted by the FIA which, at Mosley's request, included all of the contracts that Ecclestone has with the individual television companies, has been a long time coming. The FIA was also criticized in the report, as it was claimed that it had: 'abused its dominant hold over motor racing to restrict competition' (*Financial Times* 30 June 1999).

The FIA has also moved its offices from Paris to Geneva, outside the jurisdiction of the Commission.

Bernie Ecclestone's influence on global Formula 1 Grand Prix, has been at least as great as that of the 'Robber Barons' Rockerfeller, Ford, Carnegie *et al.* on the structure of America's industrial sector at the turn of the century, albeit a role played on a smaller stage. However, the authors are convinced that the world of Formula 1, and of motor racing at large, has been irrevocably changed by the business genius of Bernie Ecclestone. The extent of his influence has also been a major element in the success of the British motorsport industry. The turbulent tale of Mr Ecclestone's TV rights and consequent bond sale is continued on p. 42.

The Impact of Television

At our interview with Mosley in April 1998, he made the analogy between F1 and a TV soap that took place over a season. The characters and plot change constantly within the parameters of the overall backdrop. Motor racing sponsors are provided with a seamless season of events, exactly like the World Rally Championship, and unlike many major sports. However, those rival sports are also less complex and expensive to stage and do not involve the huge logistical effort of transporting several tons of high tech equipment across thousands of miles sixteen or seventeen times a year. Indeed, the model of F1 and the extensive reorganization of its management and presentation to the world, is one which is now being followed by other sports including football and rugby, with athletics just coming on stream.

Whilst there were many real issues of dispute between FISA and FOCA, there were also a number of issues about power; who was to run the sport? It was also apparent the sport and associated business, had to be of paramount interest to everyone involved. The money involved had developed under the management expertise of FOCA, very clearly led by Bernie Ecclestone and aided by Max Mosley. The promoters, working with the power of global television, had taken the initiative away from the international federation that had traditionally managed and organized all aspects of international motorsports.

Formula 1 Grand Prix and much of motorsport has become the joint property of the promoters and the traditional organizers, because of the impact of television money, the sponsorship of global corporations and the TV viewer, all of whom are now more important than the trackside spectator. The impact of digital television will probably bypass any of the problems created by the ban on tobacco advertising. Moreover, as is shown by the statistics presented in Tables 2.1 and 2.2, tobacco advertising was only some 8.8 per cent of the total team sponsorship in 1998, down from 12.5 per cent in 1993. This contrasted with tobacco's gain in circuit advertising in the same period when the number of companies rose from 1 to 3 and the percentage rose from 4.3 to 7.1, presumably because of differing national rules. However, tobacco companies such as Marlboro and British American Tobacco remain large players in the support of the sport.

FISA made several attempts in the early 1980s to encourage TV companies to pay for the privilege of televising the Grand Prix season. In April 1998, Max Mosley recounted the story of how Ecclestone managed to get an agreement from the European Broadcasting Union to show all of the rounds and originate the TV coverage. Therefore, there was one supplier and one fee, but the EBU's agreement was subject to certain conditions, including the

Table 2.1   **Team Sponsorship – Primary Sponsors: 1993 F1 Season**

Industry	No. of sponsors	Share %
Motor	31	55.4
Tobacco	7	12.5
Industrial	5	8.9
Clothing	4	7.1
IT	3	5.4
Drink	1	1.8
FMCG	1	1.8
Electrical	1	1.8
Photographic	1	1.8
Machinery	1	1.8
Tourism	1	1.8
Grand total	**56**	**100.0**

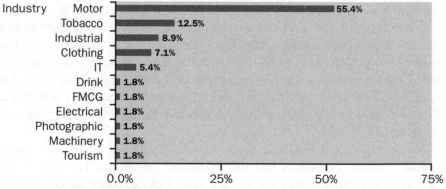

Source: Sports Marketing Surveys

proper presentation of all trackside advertising so that it did not interfere with the camera coverage. Another issue related to the French oil company Elf and the prominence of their logo at the tracks. This preference towards the French company did not endear other companies to Formula 1 on TV, nor encourage them to pay reasonable sums for appearing alongside the Elf logo. The new agreement effectively ensured that such advertising anomalies were overcome, boosting major sponsorships and advertising.

At the time of the 1981 Concorde agreement, it was known that the total television audience of Formula 1 was spread over a total of 44 countries and estimated at some 12 m; but what 12 millions? By 1987 the audience had grown and the total TV rights income was some $5 m, of which the FIA would have received approximately $1.5 m. By the beginning of the 1990s the total annual TV audience worldwide was estimated at 17.6 billion in 144 countries which had increased to more than 50 billion in 202 countries by 1999 (source

Table 2.2   Team Sponsorship – Primary Sponsors: 1998/1999 F1 Seasons

(1998)			(1999)		
Industry	*No. of sponsors*	*Share %*	*Industry*	*No. of sponsors*	*Share %*
Motor	19.0	27.9	Motor	18.0	26.5
Tobacco	6.0	8.8	Tobacco	8.0	11.8
Clothing	5.0	7.4	FMCG	7.0	10.3
Financial	4.0	5.9	Telecommunication	7.0	10.3
Drink	4.4	2.0	Financial	5.0	7.4
FMCG	3.0	4.4	IT	5.0	7.4
Computer games	3.0	4.4	Clothing	3.0	4.4
Industrial	2.0	2.9	Drink	2.0	2.9
Electrical	2.0	2.9	Industrial	2.0	2.9
Airline	2.0	2.9	TV	2.0	2.9
Mobile phone	2.0	2.9	Catering	1.0	1.5
Telecommunication	2.0	2.9	Computer games	1.0	1.5
Photographic	1.0	1.5	Courier	1.0	1.5
Ceramics	1.0	1.5	Electronics	1.0	1.5
Courier	1.0	1.5	Film	1.0	1.5
Inline skates	1.0	1.5	Inline skates	1.0	1.5
Jewellery	1.0	1.5	Stationery	1.0	1.5
Stationery	1.0	1.5	Tourism	1.0	1.5
TV 1	1.5	2.0			
Grand total	**68**	**100.0**	**Grand total**	**68**	**100.0**

Source: Sports Marketing Surveys

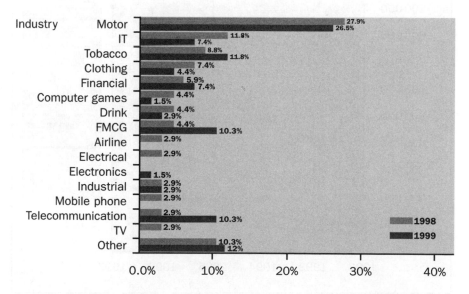

Source: Sports Marketing Surveys

FIA). This huge global audience attracts the sponsorship (advertising) spend of many companies looking for relationships that will enhance their own brand awareness. They generate exposure that cannot be achieved, except by even greater expenditure, in other ways. The growth of television audiences is shown in Tables 2.3–2.4 below.

Table 2.3   **Formula 1 Fans Tune In ... in Record Numbers**
Growth of television audience

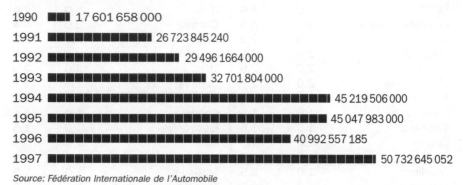

1990   17 601 658 000
1991   26 723 845 240
1992   29 496 1664 000
1993   32 701 804 000
1994   45 219 506 000
1995   45 047 983 000
1996   40 992 557 185
1997   50 732 645 052

Source: Fédération Internationale de l'Automobile

Table 2.4   **All Formula 1 Audiences by Year**

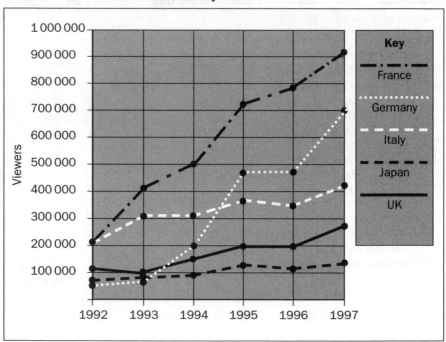

Source: Sports Marketing Surveys

Now, major companies have the opportunity to place their brand image before the largest single audience of sport in the world. Sponsors make the relationship because the image of Formula 1 is appropriate to the company's aspirations and products. Finally, the companies who pay for the sponsorship take a great deal of trouble to utilize the relationship by using the sponsorship to enhance their advertising campaigns and the image of their products. Finally, the organizations involved sometimes actually design products that carry the brand names of the teams concerned. The team names are becoming 'brand names' in their own right.

The Formula 1 organization that is Ecclestone's creation has proved a model that others, including the Touring Car promoters (TOCA), have followed. In 1998 the worldwide television audience that followed the British Touring Car Championship was nearly 2 bn and more than 1.2 bn of that number were viewing from Asia and the Pacific Rim (source TOCA). TOCA was the only major area of motor-sport that is not run by a committee of FISA, but suffered falling audiences in 1999–2000 and reorganized around cheaper technical rules for 2001.

The FIA remain responsible for the World Rally Championship and provide that global sport with the regulatory backing. The Prodrive company, run by the ex-world Rally Championship co-driver Dave Richards and his co-director Ian Parry, is an example of the WRC associations that thrive. For this is an exceptionally professional and profitable organization that has a turnover in excess of £50 m and has materially altered the fortunes of the brands it has been associated with, particularly the Subaru case examined in this book.

The marketing element of Prodrive has always been at the forefront of their philosophy. They have been so successful in the management stakes, that Dave Richards was put in charge of the Benetton Formula 1 team until it became clear that he would not be given the necessary executive freedom. This company also lives and dies by the success of their marketing, but they are slightly different from a pure race team in that they also act as automotive engineering consultants to a number of companies. They are also developing special cars that will carry the Prodrive badge in the same way that Fords carried the Cosworth badge and certain Mercedes models sport the AMG 'designer label', based on motorsport success and commercial redevelopment for the showroom. This is all about product differentiation and branding, and rallying offers unique opportunities to tie together a reputation for speed and durability in a product visibly related to the showroom offering.

Table 2.5 1999 FIA World Rally Championship Television Broadcast Statistics
Event by Event Totals

Event	Viewers	Minutes Broadcast	Broadcasting Countries
Rallye Automobile Monte Carlo	337 390 980	22 705	182
The International Swedish Rally	335 054 419	16 171	182
Safari Rally Kenya	335 944 084	15 102	182
TAP Rallye de Portugal	345 443 797	20 870	182
Rallye Catalunya Coasta Brava – Rallye de España	335 395 148	10 336	182
Rallye de France – Tour de Corse	333 107 381	24 483	182
Rally Argentina	342 138 399	17 525	182
Acropolis Rally	332 363 696	23 826	182
Rally of New Zealand	345 351 333	17 236	182
Neste Rally Finland	346 515 108	14 627	182
555 China Rally	342 695 984	14 760	182
Rallye Sanremo – Rallye d'Italie	350 145 829	21 247	182
Telstra Rally Australia	352 908 728	16 878	182
The Network Q Rally of Great Britain	358 657 807	21 318	182
Totals	**4 793 110 695**	**257 083**	**182**

Final Résumé

	Viewers	Minutes Broadcast	No. of Broadcasts	Broadcasting Countries
1999 (14 events)	4 793 110 695	257 083	13 227	182
1998 (13 events)	1 480 655 123	118 504	4 583	71
Change 1998 to 1999	+ 3 312 455 572	+ 138 579	+ 8 644	+ 111
% Change	+ 223.7	+ 116.9	+ 188.6	+156.3

Source: Sports Marketing Surveys

Table 2.6 1999 FIA World Rally Championship Television Broadcast Statistics
Country by Country

Territory	Method	Events Broadcast	Viewers	Minutes Broadcast	No. of Broadcasts
Algeria	c	1	3 072 673	220	28
Argentina	d	14	28 550 000	668	18
Australia	d	14	7 400 000	1 154	36
Austria	f	14	2 221 614	5 797	250
Bahrain	c	1	279 334	220	28
Belarus	c	14	192 836	5 656	241
Belgium	f	14	3 979 150	7 068	282

Territory	Method	Events Broadcast	Viewers	Minutes Broadcast	No. of Broadcasts
Bulgaria	c	14	418 061	6 656	241
Chile	a	14	1 440 000	350	14
Croatia	f	14	3 631 982	5 996	255
Cyprus	c	14	•	5 876	269
Czech Republic	c	14	723 133	5 656	241
Denmark	c	14	986 000	5 656	41
Egypt	c	14	•	220	28
Estonia	c	14	82 859	5 656	241
Finland	f	14	12 851 004	7 270	307
France	f	14	29 779 326	6 386	270
Germany	c	14	14 446 000	5 656	241
Greece	c	14	•	5 471	241
Hungary	c	14	1 657 181	5 656	241
Iceland	c	14	•	5 656	241
Iraq	c	14	1 939 819	220	28
Ireland, Rep. of	c	14	478 323	5 656	241
Israel	c	14	1 273 337	5 876	269
Italy	f	14	23 208 896	6 260	272
Jamaica	a	5	360 000	130	5
Japan	d	14	230 555 633	1 836	85
Jordan	c	14	•	220	28
Kenya	a	1	•	71	1
Kuwait	c	14	1 357 873	220	28
Latvia	c	14	79 093	5 656	241
Lebanon	c	14	82 962	5 876	269
Libya	c	14	•	220	28
Lithuania	c	–	97 924	5 656	241
Luxembourg	–	14	•	5 656	241
Malaysia	–	–	2 698 000	–	7
Malta	–	14	•	–	241
Moldovia	–	14	7 130 000	–	241
Monaco	–	14	7 130 000	–	19
Morocco	–	14	2 127 748	–	28
Netherlands	–	–	2 302 000	5 656	–
New Zealand	a	14	2 010 000	434	17
Norway	c	14	1 255 896	5 260	241

Territory	Method	Events Broadcast	Viewers	Minutes Broadcast	No. of Broadcasts
Oman	c	14	193 982	220	28
Paraguay	a	14	2 893 833	598	23
Peru	b	14	4 755 000	754	29
Poland	a	14	27 479 000	6 722	282
Portugal	f	14	17 700 093	6 381	275
Qatar	c	14	476 187	220	28
Romania	f	14	5 606 221	5 753	253
Russia	c	14	1 604 452	5 656	241
Saudi Arabia	c	14	5 121 122	220	28
Slovakia	c	14	677 938	5 656	241
Slovenia	c	14	271 175	5 656	241
Spain	d	14	19 523 895	8 677	319
Sweden	f	14	9 438 200	6 276	264
Switzerland	c	14	2 071 476	5 656	241
Syria	c	14	•	220	28
Tunisia	c	14	170 332	220	28
Turkey	c	14	10 130 446	6 156	283
UAE	c	14	581 946	220	28
UK	f	14	38 934 000	6 348	273
Ukraine	c	14	429 360	5 656	41
United States	b	14	7 350 000	1 092	42
Uruguay	a	14	568 500	347	14
West Bank & Gaza	c	14	•	220	28
Yemen	c	14	708 442	220	28
Yugoslavia	c	14	451 958	5 656	241
Worldwide	f	14	4 200 000 000	12 600	2 520*
Totals			**4 793 110 695**	**257 083**	**13 227**

*Rally highlight package broadcast to 180 countries – with an estimated audience of between 200–400 million per event.
• Figures not supplied by broadcaster.
~ Figure includes viewers for the south of France.
– Figure does not include viewers for all broadcasts.
a National terrestrial Free Over The Air broadcaster(s).
b Pay/Subscription broadcaster(s).
c Satellite Free/cable Free broadcaster(s).
d National Terrestrial Free Over The Air & Pay/Subscription broadcaster(s).
e National Terrestrial Free Over The Air, Satellite Free/Cable Free & Pay/Subscription broadcaster(s).
f National Terrestrial Free Over The Air & Satellite Free/Cable Free broadcaster(s).
Source: Sports Marketing Surveys

Table 2.7 1999 FIA World Rally Championship Television Broadcast Statistics:
Totals by region

Central and Eastern Europe	*Method*	*Events Broadcast*	*Viewers*	*Minutes Broadcast*	*No. of Broadcasts*
Belarus	c	14	192836	5 656	241
Bulgaria	c	14	418061	5656	241
Croatia	f	14	3631982	5996	255
Czech Republic	c	14	723133	5656	241
Estonia	c	14	82859	5656	241
Hungary	c	14	1657181	5656	241
Latvia	c	14	79093	5656	241
Lithuania	c	14	97924	5656	241
Moldovia	c	14	•	5471	241
Poland	e	14	27479000	6722	282
Romania	f	14	5606221	5753	253
Russia	c	14	1604452	5656	241
Slovakia	c	14	677938	5656	241
Slovenia	c	14	271175	5656	241
Ukraine	c	14	429360	5656	241
Yugoslavia	c	14	451958	5656	241
Sub-totals			**43403174**	**91814**	**3923**

Western Europe & Mediterranean	*Method*	*Events Broadcast*	*Viewers*	*Minutes Broadcast*	*No. of Broadcasts*
Austria	f	14	2221614	5797	250
Belgium	f	14	3979150	7068	282
Denmark	c	14	986000	5656	241
Finland	f	14	12651004	7270	307
France	f	14	29779326	6386	270
Germany	c	14	14446000	5656	241
Greece	c	14	•	5471	241
Iceland	c	14	•	5656	241
Ireland, Rep. of	c	14	478323	5656	241
Italy	f	14	23208896	6260	272
Luxembourg	c	14	•	5656	241
Malta	c	14	•	5656	241
Monaco	c	14	7130000▲	443	19
Netherlands	c	14	2302000	5656	241

Western Europe & Mediterranean (cont.)	Method	Events Broadcast	Viewers	Minutes Broadcast	No. of Broadcasts
Norway	c	14	1 255 896	5 760	241
Portugal	f	14	17 700 093	6 381	275
Spain	d			8 677	319
Sweden	f	14			264
Switzerland			2 071 476	5 656	
UK				6 348	
Sub-totals				**117 385**	

Far East & Pacific Rim	Method	Events Broadcast	Viewers	Minutes Broadcast	No. of Broadcasts
Australia	d	14	7 400 000	1 154	36
China	a	6	37 504 500	364	14
Japan	a	14	230 555 633	1 836	85
Malaysia	a	7	2 698 000	182	7
New Zealand	a	14	2 010 000	434	17
Sub-totals			**280 168 133**	**3 970**	**159**

North Africa & Middle East	Method	Events Broadcast	Viewers	Minutes Broadcast	No. of Broadcasts
Algeria	c	14	3 072 673	220	28
Bahrain	c	14	279 334	220	28
Cyprus	c	14	•	5 876	269
Egypt	c	14	•	220	28
Iraq	c	14	1 939 819	220	28
Israel	c	14	1 273 337	5 876	269
Jordan	c	14	•	220	28
Kenya	a	1	•	71	1
Kuwait	c	14	1 357 873	220	28
Lebanon	c	14	82 962	5 876	269
Libya	c	14	•	220	28
Morocco	c	14	12 127 748	220	28
Oman	c	14	193 982	220	28
Qatar	c	14	476 187	220	28
Saudi Arabia	c	14	5 121 122	220	28

North Africa & Middle East (cont.)	Method	Events Broadcast	Viewers	Minutes Broadcast	No. of Broadcasts
Syria	c	14	•	220	28
Tunisia	c	14	170 332	220	28
Turkey	c	14	10 130 446	6 156	283
UAE	c	14	581 946	220	28
West Bank & Gaza	c	14	•	220	28
Yemen	c	14	708 422	220	28
Sub-totals			37 516 182	27 375	1 539

The Americas	Method	Events Broadcast	Viewers	Minutes Broadcast	No. of Broadcasts
Argentina	d	14	28 550 000	688	18
Chile	a	14	1 440 000	350	14
Jamaica	a	5	360 000	130	5
Paraguay	a	14	2 893 833	598	23
Peru	b	14	4 755 000	754	29
United States	b	14	7 350 000	1 092	42
Uruguay	a	14	568 500	347	14
Sub-totals			**45 917 333**	**3 939**	**145**

Worldwide	Method	Events Broadcast	Viewers	Minutes Broadcast	No. of Broadcasts
Worldwide	f	14	4 200 000 000	12 600	2 520
Sub-totals			4 200 000 000	12 600	2 520
TOTAL			**4 793 110 695**	**257 083**	**13 227**

- • Figures not supplied by broadcaster.
- ~ Figure includes viewers for the south of France.
- − Figure does not include viewers for all broadcasts.
- a National terrestrial Free Over The Air broadcaster(s).
- b Pay/Subscription broadcaster(s).
- c Satellite Free/cable Free broadcaster(s).
- d National TerrestrialFree Over The Air & Pay/Subscription broadcaster(s).
- e National Terrestrial free Over The Air, Satellite Free/Cable Free & Pay/Subsscription broadcaster(s).
- f National Terrestrial Free Over The Air & Satellite Free/Cable Free broadcaster(s).

Source: Sports Marketing Surveys

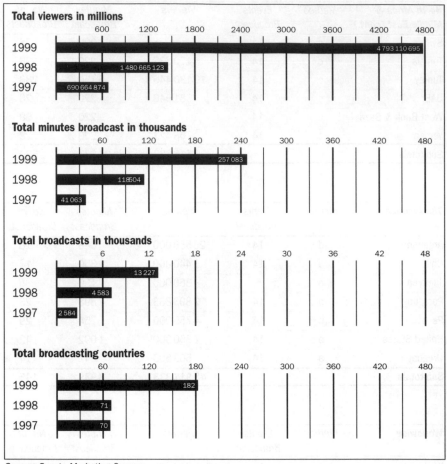

Source: Sports Marketing Surveys

Fig. 2.1   **FIA World Rally Championship: Television broadcasting statistics – live and highlights coverage (not including news broadcasts)**

All of these companies and teams that are involved in the sponsorship of motorsports, have one thing in common; the reach of global television increasingly defines their market place. Tables 2.8–2.10 detail the sponsors by team and by circuits in the seasons 1993 and 1998 respectively. The data clearly demonstrates that the spread of companies has increased between 1993 and 1998, particularly as far as circuit advertising is concerned and the areas from which companies come to support F1 teams. They are no longer confined to tobacco companies; two teams in the 1999 season had no tobacco sponsorship. However, circuit advertisers recorded gains in both the number of tobacco companies participating and their percentage of total circuit advertising (see Tables 2.11 and 2.12).

Table 2.8 **1993 Sponsors by team**

Benetton	BENETTON	Clothing
	CAMEL	Tobacco
	ELF	Motor
	FORD	Motor
	SULA SUGARFREE	FMCG
BMS	AGIP	Motor
	CHESTERFIELD	Tobacco
Ferrari	AGIP	Motor
	FIAT	Motor
	MAGNETI MARELLI	Motor
	MARLBORO	Tobacco
	PIONEER	IT
Footwork	BP	Motor
	FOOTWORK	Motor
	JAPAN	Tourism
	TOSHIBA	IT
Jordan	ARISCO	Motor
	DIAVIA	Motor
	PERAR (Ball valves)	Motor
	SASOL	Motor
	UNIPART	Motor
Larrousse	CHRYSLER	Motor
	ELF	Motor
	LARROUSSE	Motor
	ZANUSSI	Electrical
Ligier	ELF	Motor
	GITANES BLONDES	Tobacco
	MAGNETI MARELLI	Motor
	RENAULT	Motor
Lotus	CASTROL	Motor
	HITACHI	Machinery
	KOMAT'SU	Industrial
	LOCTITE	Industrial
	SHIONOGI	Industrial
McLaren	BOSS	Clothing
	COURTAULDS	Clothing
	KENWOOD	IT
	MARLBORO	Tobacco
	SHELL	Motor
Minardi	AGIP	Motor
	BETA	Motor
	COCIF	Industrial
	MERCATONE	Industrial
	MINARDI	Motor
Sauber	ELF	Motor
	LIQUI MOLY	Motor
	MERCEDES	Motor
Tyrrell	BP	Motor
	CABIN	Tobacco
	CALBEE	Clothing
	YAMAHA	Motor
Williams	CAMEL	Tobacco
	CANON	Photographic
	ELF	Motor
	LABATT'S	Drink
	RENAULT	Motor

Source: Sports Marketing Surveys

Table 2.9   **1998 Sponsors by team**

Arrows	DANKA	IT
	IXION	IT
	PARMALAT	FMCG
	ZEPTER	IT
Benetton	AGIP	Motor
	AKAI	Electrical
	BENETTON	Clothing
	D2	Mobile phone
	FEDEX	Courier
	KOREAN AIR	Airline
	MILD SEVEN	Tobacco
	PI.SA	Ceramics
Ferrari	ASPREY	Jewelery
	FIAT	Motor
	MAGNETI MARELLI	Motor
	MARLBORO	Tobacco
	SHELL	Motor
	TELECOM ITALIA	Telecommunication
Jordan	BENSON & HEDGES	Tobacco
	G DE Z	Financial
	NATWEST	Financial
	PEARL	Financial
	PLAYSTATION	Computer games
	REPSOL	Motor
	S. OLIVER	Clothing
McLaren	BOSS	Clothing
	computer associates	IT
	FINLANDIA	Drink
	KENWOOD	IT
	LOCTITE	Industrial
	MERCEDES	Motor
	MOBIL 1	Motor
	SCHWEPPES	FMCG
	SIEMENS	IT
	SUN	IT
	WARSTEINER	Drink
	WEST	Tobacco
Minardi	FONDMETAL	Motor
	MTA	Electrical
	ROCES	Inline skates
Prost	AGFA	Photographic
	ALCATEL	Mobile phone
	BIC	Stationery
	CANAL+	TV
	GAULOISES	Tobacco
	PEUGEOT	Motor
	PLAYSTATION	Computer games
	TOTAL	Motor
Sauber	PETRONAS	Motor
	PLAYSTATION	Computer games
	RED BULL	FMCG
Stewart	FORD	Motor
	HSBC	Financial
	LEAR	Aeronautical
	MCI	Telecommunication
	TEXACO	Motor
Tyrrell	BROTHER	Industrial
	EUROPEAN	Airline
	FORD	Motor
	MORSE	IT
	PIAA	Motor
	YKK	Clothing
Williams	AP LOCKHEED	Motor
	CASTROL	Motor
	FALKE	Clothing
	SONAX	Motor
	VELTINS	Drink
	WINFIELD	Tobacco

Source: Sports Marketing Surveys

Table 2.10   **Circuit advertisers: 1993 F1 season**

Industry	No. of sponsors	Share %
Motor	10	43.5
Drink	3	13.0
IT	3	13.0
Tourism	2	8.7
Electrical	1	4.3
Newspaper	1	4.3
Ski equipment	1	4.3
Tobacco	1	4.3
White goods	1	4.3
Grand total	23	100.0

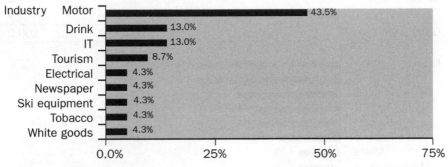

Source: Sports Marketing Surveys

The markets that these companies/sponsors were and are fighting for are global and multi-cultural, transcending national and regional boundaries. Companies provide the sponsorship money to the constructors, but television (read ITV for Formula 1 live broadcasts in Britain) provides more money; television advertising is the main incentive for the sponsors. Each of the Formula 1 constructors who are contracted to the circus, represent a regular hardcore of points scoring competitors, and for that they receive a guaranteed income of between £5.8 and £14.4m per season. However, it has taken Bernie Ecclestone and the Formula 1 Constructors Association more than a quarter of a century to achieve such relative financial stability.

TV income is additional to the sponsors that individual teams attract. It is the advertising income, or a part of it, that the TV companies use to fuel the motorsport spectacle. More particularly the advertisers are looking for specific audiences in terms of age, socio-economic group and gender. Whilst the popularity of motorsport has grown around the world the profile of the TV audience has changed. The Formula 1 audience in particular is no longer simply the 'petrolheads'; it has become one of the most glamorous and prestigious sports in the calendar.

Table 2.11 **Circuit advertisers: 1998 F1 season**

Industry	No. of sponsors	Share %
Motor	10	23.8
Drink	7	16.7
Financial	5	11.9
IT	4	9.5
Tourism	4	9.5
Tobacco	3	7.1
Airline	2	4.8
Mobile phones	2	4.8
Coffee	1	2.4
FMCG	1	2.4
Media	1	2.4
Supermarket	1	2.4
TV	1	2.4
Grand total	42	100.0

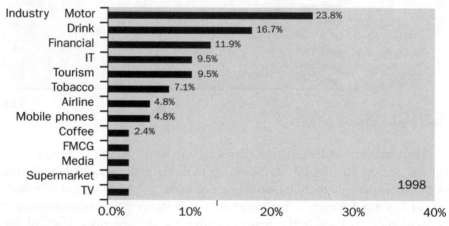

Source: Sports Marketing Surveys

Advertisers are looking for sports/attractions that provide them with an audience of customers or potential customers. In Western Europe football and F1 audiences have a similar male/female split, with F1 audiences 63/37 per cent, but this has altered significantly during the past decade from a split of 80/20 per cent. However F1 audiences have a significantly higher socio-economic profile and are slightly older, than for football. Whilst athletics and ice-skating have the highest levels of female viewers, the audience for figure skating is significantly older. Golf is the one sport that has a higher socio-economic profile than motorsport but the audience tends to be significantly older and smaller; see Tables 2.13 and 2.14. Table 2.13 shows that to some extent the popularity of the sport feeds upon itself; as the audience grows, so does the coverage.

Table 2.12   **Demographic profile of F1 audience**

Demographic profile	F1 TV audience	% of population	Deviation: F1/Pop. %
Male	63	49	+27.8
Female	37	51	–27.0
Age 4–15	10	21	–54.2
Age 16–24	8	11	–22.7
Age 25–34	18	16	+15.5
Age 35–44	17	14	+22.6
Age 45–54	17	13	+27.5
Age 55–64	13	10	+34.0
Age 65+	16	15	+8.3

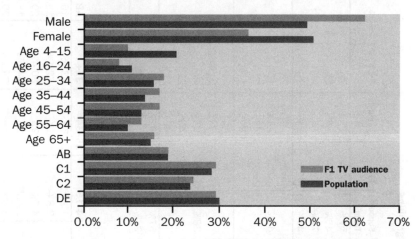

Source: Sports Marketing Surveys

Television's Continued Importance

Such televisual economic attractions are traded and, like any other valuable commodity, the price depends on the scarcity and the urgency of demand. Hence, the huge stream of TV and advertising cash that has rolled into F1. Nevertheless, the TV companies are not immune from economic forces and a momentary hesitation can be disastrous. Just such a salutary lesson was handed out to the British Broadcasting Corporation at the end of the 1995 season.

Table 2.13   **Formula 1 TV Viewer Demographics**

	Viewed 10+OP's	Support Team	Men	Women	6-14	15-34	38-49	50+	Upper	Middle	Lower
Western Europe											
France	33%	57%	61%	39%	7%	22%	22%	49%	39%	41%	20%
Germany	47%	46%	65%	35%	5%	35%	21%	40%	51%	38%	11%
Italy	28%	72%	66%	34%	6%	31%	30%	33%	34%	49%	17%
UK	54%	41%	62%	38%	9%	27%	25%	39%	46%	24%	30%
Central Europe											
Czech. Rep	38%		69%	31%	9%	24%	27%	40%	11%	22%	67%
Hungary		44%	59%	41%	10%	27%	24%	39%	11%	26%	66%
Poland			58%	42%	12%	35%	34%	19%	36%	35%	29%
Russia			71%	29%	11%	28%	26%	35%			
Asia											
China			58%	42%	17%	39%	25%	19%			
Indonesia			57%	43%	21%	39%	20%	20%			
Japan	19%	38%	66%	34%	4%	47%	35%	14%			
Philippines			57%	43%	21%	44%	19%	16%			
Thailand			55%	45%	22%	44%	19%	15%			
South America											
Argentina	18%	67%	58%	42%	9%	31%	28%	32%	19%	21%	66%
Mexico	15%	27%	57%	43%	23%	42%	20%	15%			

Other Sport Viewer Demographics

Sporting Events	Men	Woman	4-15	16-34	35+	AB	C1	C2	DE
Football	62.9%	37.1%	10.6%	25.8%	63.6%	16.9%	26.6%	23.7%	32.8%
Athletics	53.2%	46.8%	8.8%	19.1%	72.1%	18.6%	26.6%	22.4%	32.4%
Golf	59.0%	41.0%	4.7%	14.9%	80.4%	20.5%	28.2%	21.3%	30.0%
Ice Hockey	53.7%	46.3%	12.5%	33.0%	54.5%	18.4%	20.4%	40.8%	20.4%
Skiing	51.0%	49.0%	8.6%	17.2%	74.2%	20.7%	27.5%	19.8%	32.1%
Figure Skating	44.4%	55.6%	7.5%	12.8%	79.6%	19.2%	28.8%	19.6%	32.4%

Source: Sports Marketing Surveys

Table 2.14   Formula 1 – The Growth: TV Coverage

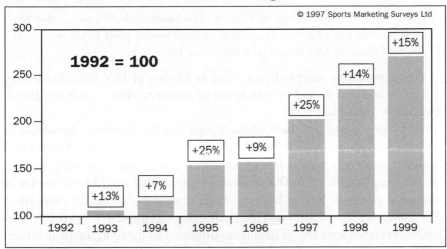

Source: Sports Marketing Surveys

Table 2.15   Formula 1 – The Growth: TV Audience

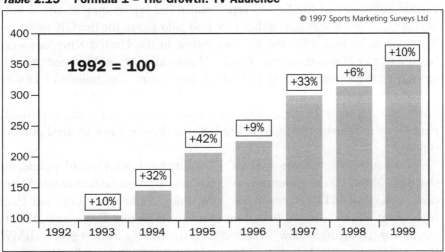

Source: Sports Marketing Surveys

The BBC had literally been the one company that had carefully nurtured the televising of the fledgling Grand Prix spectacle post-war, and one man had been responsible for such dedication continuing in the 1980s and after, one Jonathan Martin. But on 13 December 1995 he received a terse telephone call from Bernie Ecclestone. Ecclestone told Martin that he had sold the Television UK broadcasting rights of the Formula 1 circus to Independent Television for £65 m. Mr Martin reportedly gave the predictable response: 'Bernie, it would have been nice to have been able to put in a competitive bid.'

Ecclestone's response is not so predictable:

Unless you have been cheating me all these years Jonathan, there is no way that you could possibly pay what they are paying, so there was no point in talking to you about it. (*Financial Mail* on Sunday 1 November 1999).

That report also observed that Andrew Chown of ITV said: 'Normally I deal with committees and various levels of authority. With F1 it is just Bernie Ecclestone – full stop.'

That report illustrates the continued trust that the Formula 1 constructors, and the Federation Internationale de L'Automobile, have for Ecclestone's business acumen. Secondly, it is a classic example of the negotiating skill and blunt honesty of the man. Thirdly, it also suggests that the FIA's desire for an executive structure for the business Ecclestone runs has yet to be fulfilled. This concern endures, although Mr Ecclestone was quoted in April 2001 with tongue in cheek sincerity as having no intention of dying in the near future.

The most fundamental change in motorsport in the past decade has been the rise in the value of the TV rights, and it is increasingly difficult to be certain about any of the figures reported. In the report quoted above on 1 November 1999, it was stated that ITV had paid £65 m for the UK rights for 5 years. On 26 June 1999 the viewing public in the United Kingdom were treated to an old movie starring Goldie Hawn rather than the timed practice session of the French Grand Prix at Magny Cours. The failure of ITV to televise the spectacle led to reports that the dispute with the 'bosses of Formula 1' (*Sunday Times*, 27 June 1999) was to blame. ITV had reportedly paid £80 m to televise the races in the UK, but they had not secured practice rights.

As at April 2001, the sale of SLEC F1 equity has been a turbulent tale, the share of holdings then amounting to 25 per cent Ecclestone family, and 51 per cent to Kirch-EMTV German TV companies. Although Kirch and their rescuers at EMTV wanted to sell on their assets to pay-TV sources, the six major car companies now involved in Grand Prix (Fiat-Ferrari, BMW, Renault, Ford-Jaguar, Honda and Mercedes) opposed this totally on grounds of the danger of pay-TV providing only a restricted audience. In May 2001 EMTV confirmed losses of 1.4 billion euros for 2000. Their share values plummeted 94 per cent. EMTV founder Thomas Haffa resigned, ironically replaced by BMW heiress Susanna Quandt's husband, Werner Khattern. The saga continues ...

By 24 April 2001, an apparent 100 year deal had been finalised. To quote from *Autosport* of that date:

FIA confirms 100-year F1 commercial agreement

Motorsport's governing body, the FIA, has confirmed its 100 year deal for the commercial rights of Formula 1 with Grand Prix ringmaster Bernie Ecclestone and his family trust company, SLEC. At an extraordinary meeting at Paris today (Tuesday April 24, 2001), agreement was reached between the FIA and SLEC, represented by Kirch boss Thomas Haffa and Dieter Hahn of EMTV. Plus the Formula 1 Association and Formula 1 Management, both represented by Ecclestone.

German media companies Kirch and EMTV jointly own 75 percent of SLEC, while Ecclestone holds the remaining 25 percent. The new deal begins when the current contract, also with SLEC, expires at the end of 2010. The deal was always likely to be rubber stamped as the FIA must comply with the European Commission's requirement to separate its commercial and promotional activities from the sport. The deal is for such a long period to avoid the likelihood of repetitive wrangles over competition issues in the future. SLEC now holds the commercial and broadcast rights to F1 until the next century, and both it and the Kirch Group have confirmed the championship will continue to be shown on free-to-air television, rather than exclusively on its digital pay-per-view channels.

The five European based F1 manufacturers Ford (Jaguar), Fiat (Ferrari), Mercedes-Benz, BMW and Renault are unhappy that the German media companies have been allowed to gain control of the commercial aspect of the sport, and have threatened to set up a rival series when the Concorde Agreement ends in 2008. Kirch indicated over the weekend that a sale of a percentage of its share to the manufacturers, thought to be as much as 45 percent, is imminent.

The *Autosport* report concluded: 'Doubtless we have only seen the start of such TV wars as the sums of money and potential audiences are so large.'

Ecclestone himself has backed the TV future, by investing phenomenal amounts in the new digital TV centre that is working now, but hidden on terrestrial TV for views in ITV Britain 2001. Ecclestone's investment is reputed to have been up to $100 m and the yearly running costs of the set up is estimated at $20 m. These figures appear vast, but so is the enterprise that includes several large transporters and two Boeing 747 aeroplanes to transport them to the races around the world.

However, the rewards are commensurate. *Autosport* online reported on 23 April 2001:

Bernie Ecclestone has taken the top slot in this year's biggest risers section of *The Sunday Times* Rich List, following the sale of a further 25 percent of his Formula 1 business earlier this year. The purchase of a further quarter of SLEC, Ecclestone's holding company which owns the commercial and broadcasting rights to F1, by German media group Kirch, added to the worth of his remaining 25 percent stake has increased Ecclestone's wealth by an estimated £1000m, a third of his overall wealth.

Bernie Ecclestone was third in the overall UK Rich List 2001, having a fortune estimated at £3bn.

The future of the FIA is now secure, as Mosley and the FIA Senate with Balestre as Chairman, have limited the period in which Ecclestone's company has the rights. Britain has retained the technological and managerial lead in this multi-billion pound global business, but it would be appalling if British antipathy towards the creation of new wealth led to our future failure in this particular area, as we have failed in so many other commercial arenas of the 20th century.

PART II

THE STORY OF TWO CAMPAIGNS: THE SUBARU STORY

The Reshaping of a Brand: Subaru and the World Rally Championship

3

Fuji Heavy Industries and Subaru in the 1980s

The Problem

There is one Japanese automobile manufacturer that has created a totally new marketing image and a new brand, within a period of twelve years. Of all the Japanese manufacturers in the early post-war years (founded in 1953), Fuji Heavy Industries was one with high-technology antecedents. The company can trace its origins back to the second decade of the 20th century, when the company was founded as an aeronautical research laboratory. The parent company Fuji Heavy Industries has always been a technological leader, it is one of the world's top five producers of general purpose engines, designing and manufacturing engines for a wide variety of uses, including marine leisure craft, road cars, trucks, construction vehicles and equipment. FHI is also an important sub-contractor to the aerospace industry, building components for military and commercial aircraft as well as un-manned aircraft. Active in the Japanese space programme, the company's technological expertise was used to design and build the Lunar Lander FTB (Flying Test Bed) for the Japanese Space Agency. The company has five main divisions, with total turnover in 1999 of over £8098 m, with more than 80 per cent of turnover (over £7170 m) derived from the automobile division.

The company launched its first motor car in 1958, which was a mini car with a 360cc engine. Then in May 1966, the company launched the first mass-produced front wheel drive vehicle in Japan. This car had a flat four, horizontally opposed engine of a similar layout to the engine designed by Professor Ferdinand Porsche for the Volkswagen car of 1936. The technological developments continued with the launch of the first all-wheel drive vehicle in 1980, a design which has since been the basis for all Subaru drive systems. At least as important as the all-wheel drive system was the new power unit that was a development of the original 'boxer' engine with four horizontally opposed cylinders.

The company has a long history of producing mini commercial vehicles for the Japanese market and throughout the 1970s and early 1980s the company continued with its mini-vehicle dominated product line. By 1986 the company was producing a total of 238 992 mini vehicles against some 71 417 cars. Whilst the company was profitable, the dominance of the mini-vehicles (cars and commercial vehicles) severely limited potential profitability and opportunities to develop new products. Output of mini vehicles accounted for over 70 per cent of total output and these cars and vehicles were produced to a specification that had an advertising slogan of 'inexpensive and built to stay that way'.

Both Mr Kuze (ex-President of STI) and Mr Yamada (current President of STI) have emphasized that the company's development programme had been clearly defined since 1984, some four years before the launch of the Legacy and the new Boxer engine. This programme was expressed as follows:

• Redesign the horizontally opposed engine to improve performance
• Develop a new vehicle with all wheel drive and controllability
• Concentrate upon the 3S's of Speed, Style and Space
• Develop three main models based upon the same platform
• Use motorsport to establish a distinctive set of brand values and image

The company pursued these strategies with deliberate consistency. The Subaru all-wheel drive system is unique in that it has a central gearbox and is capable of being powered by relatively small engine units. The layout offers a number of advantages over other systems, including the flexibility of the boxer engine design, which gives a wide range of performance options from 1.5 to 2.5 litres. Also, the engine's low profile offers a low centre of gravity and allows the chassis and suspension design to be simplified. In addition to the considerable cost of these technological advances the company designed a completely new passenger car, the Legacy, which would incorporate all of the advances.

However, as Mr Kuze, the past President of Subaru Tecnica International pointed out in an interview in December 1999, these new developments of the mid-1980s meant that the company had already 'bet its future'. In particular the launch of the new 'boxer' engine in 1989, along with the development of a new facility to build the engines, required an investment of some $160 m in 1988, which was spent at the same time as the costs of developing the Legacy. Given that the total turnover of Fuji Heavy Industries at the time was some $5 bn, it was a large and heavy investment. As Mr Yamada, then in charge of engine development, said in an interview in December 1999: 'Power is vital in a good car. Power can compensate for deficiencies in chassis design but a brilliant chassis will not compensate for a bad engine.' The fact remained that the company had to develop another series of concepts and strategies to optimize the returns from these developments. The question was how?

In spite of the huge investments in new technology, models and engines by the end of the 1980s there was no clear strategy for pursuing the development of a new brand image. The company had the base for developing new products, but around the world its products were either regarded as basic farm vehicles or relatively unknown. In Europe and the United Sates, markets where the company had to achieve a strong position if it was to justify its investments, the brand was relatively unknown. Furthermore, the proportion of mini cars and vans being sold in the domestic market remained at over 70 per cent and overseas sales and revenues were declining, as the model range failed to attract new consumers. As Tables 3.1 and 3.2 demonstrate, the company's overall financial performance was deteriorating.

Table 3.1 **Sales of Mini and Other Cars 1988–99**

Domestic Market

	1988	1989	1990
Mini passenger car	143 672	125 805	126 284
Mini truck/van	114 862	99 257	126 023
'Legacy'	–	46 724	57 542
'Impreza'	–	–	–
'Forester'	–	–	–
Others	69 458	41 092	27 436
Total	**327 992**	**312 878**	**337 285**

Overseas Market (Export + Local Production)

	1988	1989	1990
Mini passenger car	9 071	6 826	6 778
Mini truck/van	–	–	–
'Legacy'	–	98 360	110 432
'Impreza'	–	–	–
'Forester'	–	–	–
Others	263 137	125 797	119 515
Total	272 208	230 983	236 725

Table 3.2 **Company Financial Details 1988 to 1990**

	Automobile Sales (¥m)	Gross Profits (¥m)	$/¥ Annual Average
1987	603 600	105 100	159.88
1988	577 557	106 889	138.37
1989	538 795	93 160	128.31
1990	524 400	75 730	142.85

At the end of the 1980s the company's results were not encouraging and after four years of declining revenues, the President and Chairman of Fuji Heavy Industries, Toshihiro Tajima made a number of observations in the company's 1989 annual report. Firstly, he emphasized the importance of the Legacy which 'represents a new image for FHI, one which goes beyond specialized markets to appeal to a wide range of consumers'. He also emphasized the vital importance of continued investment and the future of the main markets in which the company operated. The company's overseas activities were assuming increasing strategic focus in the future of the company; as the annual report states: 'Our overseas production bases are coming on stream and our marketing structure in the United States has been considerably strengthened'. But the financial facts suggest a company that was waiting for the investments to pay off, as the total sales of the automobile division and other divisions indicate in Tables 3.1 and 3.2. The automotive division accounted for 84.3 per cent of sales in 1987 and fell to 81.2 per cent in 1989 and 79.7 per cent in 1990. However, total automotive sales had fallen –7.6 per cent in 1987, –4.3 per cent in 1988, –6.7 per cent in 1989 and –2.7 per cent in 1990. (See Table 3.2).

The Solution: Identifying the Marketing Strategy

By 1984 some of the company's key managers, including Mr Kuze, then Manager of the FHI Proving Ground (Mr Koseki was then Chief Test Driver) and Mr Narita, General Manager of Advertising, Fuji Heavy Industries, were determined that the way forward was through promotion of the company's products through motorsports. This would allow them to capitalize on the technological developments and build a unique brand image. But the path to that decision had not been smooth.

In the mid-1980s the Board had been persuaded that the advantages offered by entry into Formula 1 would outweigh the costs involved and an Italian company was commissioned to design a 12 cylinder horizontally opposed engine, with another company taking responsibility for the chassis. The argument was a persuasive one, since the successful development of such an engine would have enabled the company to enter Formula 1 as an engine supplier. This was a logical activity for one of the world's largest engine suppliers, as there would be opportunities for technological transfer as well as considerable promotional benefits. However, Fuji Heavy Industries was to find, as so many automotive companies had before them (and no doubt more will do so in the future), that success in motorsport does not come easily or cheaply; certainly not without total commitment and focus. Furthermore, gaining the necessary experience in one of the most aggressively competitive and costly activities in the world, can be a prohibitively expensive and painful process. So it was to be, and the company was to learn the hard way that building good, even excellent road cars, is not a sufficient condition for success in motorsport at international level.

This was a salutary lesson for the company, which was to define their future approach to motorsports. There is a dominant need for trust in any company or group chosen as a partner and for the partner to deliver any promised performance, or exceed such promises. That there will be problems in any such partnership is perhaps inevitable, but partners do have to be capable of working together in an atmosphere of collaboration, not of blame. In an interview with the author in December 1999, Mr Yamada emphasized these facts and the influence they had on him in his role as head of the company's motorsports. The observations were not only about engineering but the 'softer' aspects of managing joint projects, all of which were a direct result of his experience in the United States during his investigations for the development programme of the new Boxer engine.

A major organizational breakthough was achieved early in 1988 when Mr Narita and his colleague proposed that the company establish a separate research and development and motorsport division. The philosophy behind the proposal centred on the importance of concentrating on further development of the basic layout that they had achieved with the new Legacy. The Legacy was to be the first of a series of models with the same floor plan, and vehicle developments needed to be concentrated, if the company was to change their product line quickly. The recommendation to the Board to establish Subaru Tecnica International was approved on 2 April 1988 and Mr Kuze, who had been a member of the Board of FHI, was appointed the first President of STI and a main Board Director of FHI. It was also agreed that the company had to develop a worldwide market, rather than simply attempt to retain its position in the main Japanese market. It was therefore essential that the company's motorsport activities should be used as a marketing tool to achieve this objective, and Mr Kuze determined that their motorsport efforts had to be at world level.

The next major enterprise was the company's attempt on the FIA World 100 000 km Speed Record with a Legacy saloon and the new Boxer engine. The challenge was mounted in January 1989 under the control of Mr Koseki, and after 19 days and nights the car had broken the record at an average speed of 223.345 km/h. A dramatically successful result, but there were problems in using the achievement for the purposes of international marketing. Kuze's evaluation of the event was conclusive. The success was not a suitable topic for the promotion of the product over a long period and would not therefore provide a consistency of message to consumers: a sales person cannot talk about one event in 1989 forever!

Through the efforts of Mr Kuze and Mr Narita, the company had provided limited assistance to a team led by Mr Koseki that participated in the Safari Rally of 1983, with a team of three cars. This activity continued with limited success throughout the 1980s, although the lack of early success in the Formula 1 projects did not encourage the commitment of a great deal of further resources, nor did the company's financial performance. Nevertheless, with the commitment of the Board of Directors the company decided to pursue the idea of generating further motorsport activities through the World Rally Championship, led by Mr Kuze as the project leader. A major element in this decision was that by then the company had the 'longest experience of any automotive company in the world in all-wheel drive technology' and knowledge of turbocharging; both were, and are, keys to international rally success. Furthermore, rallying was, in the light of their experience, considered to be a more relevant and affordable motorsport arena to further develop the company's profile to the consumer. Both Mr Kuze and Mr Narita were very clear about the objectives that were set for the company's entry into the World Rally Championship in the late 1980s. These were as follows:

- To provide the company with an advertising opportunity to place the characteristics of the company's products before current and potential owners.
- To provide an additional opportunity and platform for the technical development of the company's products.
- To differentiate the products through distinctive branding and imaging in domestic and export markets. Subaru participated in the Group N rally series to give the cars homologation.

As the manager responsible for motorsport and President of STI, Mr Kuze had already determined that the World Rally Championship would be a more appropriate activity than track racing. The primary reasons for choosing rallying were that it provided the company with the opportunity to utilize a spectacle with 'a monthly changing subject, which gives a pride and self-confidence to Subaru customers, Subaru salesmen and Subaru workers' (Mr Kuze, December 1999). In addition there were a number of very practical aspects of rallying that had a great deal of appeal to Mr Narita and his sales and marketing teams:

- Rallying was more directly related to selling cars rather than a brand image, which is probably what Formula 1 achieves

- Rallying would provide a direct link with the values of the Subaru brand and the customer's vehicle
- Rallying provided the opportunity to use a road car and develop a more exclusive range

In the following months Mr Kuze visited the WRC events, and in April he accompanied a representative from the Royal Automobile Club's Motorsport Committee to the Safari Rally making valuable contact with Mr David Richards of Prodrive. After these developments he put forward the concept to a meeting with the European Importers and Dealers representing Subaru. They agreed with the proposal and negotiations were started with Prodrive. It was decided that Mr Yamada would take control of the engine development project but, early on in the discussions, he had emphasized his complete lack of knowledge of the rallying world. Consequently, it was agreed that he would spend the 1989 rally season following the team's efforts in running the BMW rally team, with the agreement of the Bavarian automobile manufacturer. It was an experience that proved crucial to the future development of the relationships between the two prospective partners.

During the author's interview with Mr Yamada, he provided a number of insights into his views on partnering and collaborating with other companies. He explained that during his period in the United States investigating the Boxer engine development, he had visited and observed a total of ten Japanese companies operating in America in partnership with other companies. Of those partnerships that were not working effectively, he reasoned that there were two fundamental causes of failure: 'either the American partner had forced the Japanese company to work the American way, or the Japanese company had forced the American partner to work the Japanese way'.

As he subsequently emphasized, the secret is that partnerships need to be undertaken in an atmosphere of collaboration and co-operation, not domination. It is that approach which appears to have remained the basis for the relationship that was formed between the two organizations.

At the end of the 1980s the company still had a product range and sales that were dominated by mini vehicles, and its financial position was still plagued by the low profitability per unit produced, typified in a high volume (but not high enough) low value added output. Strategically, the management realized that the next decade would be the period when the proof of whether their vision and commitment to change was an appropriate strategy. Early in the year the Board decided that it would be appropriate for new thinking to be applied. A new President and Chairman of the company, Isamu Kawai, was appointed on 28 June 1990. The new President gave his support to the plans that had been set down in the latter half of the decade and approved by the previous President. In the 1990 Annual Report, he set down the three main principles of his management philosophy:

- 'First, that the basis of all decisions is thoroughly understanding the situation and taking all factors into consideration;
- Second, that management must be from the heart if it is to motivate people;
- And third, that the principle 'the customer comes first' is fundamental to success'.

One of the major factors in the management changes was the large drop in overseas sales; they fell by 15.8 per cent in the fiscal year 1989–90 and it was obvious that a restructuring of certain overseas operations was required. It is clear that by 1990, Fuji Heavy Industries was a company that needed to change and it had, by this period, already set in place the strategy for the future. That vision and the company strategies and policies necessary to make that vision a reality were in fact already in place, it certainly included the use of motorsport sponsorship as a key element, perhaps the corner-stone of their marketing strategy. As Mr Narita stated when asked why they chose motorsport as the main factor in their marketing and sales strategy: 'Firstly, motorsport in general is a financially efficient method in marketing, but I felt that rallying was more directly related to the customers' experience on the road than Formula 1.'

The extent to which the company was 'betting its future' on the investments made in the mid-1980s was very evident to Mr Kuze at the time. In his view, the main pillars of the strategy that was laid down at that time remained appropriate now nearly fifteen years later. As he expressed it in December 1999, the main elements of their strategy were:

'Subaru is a small independent niche player and we needed a unique competitive advantage. We had developed this road car layout in the Legacy, with the Boxer engine, a unique all wheel drive system and a stiff chassis. So we had to stake the future on the new model including the new engine plant, and launch into the World Rally Championship.'

The main reasons for choosing the WRC were also very clear, and not unrelated to the expensive failure of the company's venture into Formula 1. They were summarized by Mr Yamada for the author as follows:

- World-wide coverage and TV exposure of 14 events
- Competition with a production model

- Less than one-quarter the cost of Formula 1, maximum exposure for minimum cost
- It highlighted one of the three models (all based on the same platform)
- Advantages perceived in all three models
- The WRC effect highlighted the driving qualities of the car/platform/engines
- Maximum cost effect for a small manufacturer
- Borderless appeal across cultures.

The company's thinking went considerably further and the concepts of the brand were well defined in several internal company reports during 1988–89. The detailed brand qualities and main objectives of the campaign for the brand were established as follows:

THE AIMS AND TASK IN WRC COMPETITION

To prove the excellent basic engine layout in driving performance
To show continuous challenge to appeal to the public with Winning.

Source: STI

The same report also identified the characteristics of their preferred target consumer who is defined as: 'instinctively aggressive and competitive and wanting a product that appeals to their instinct'. 'The cars offer SPEED, STYLE, SPACE'.

Mr Yamada was very clear that Subaru had to a build new brand image, based upon the characteristics and values embodied in the new passenger cars. These were totally different from the values and characteristics of the previous range of cars and mini vehicles.

However, the company had a very limited knowledge of motorsport and Yamada himself had made it very clear to Mr Kuze that he had 'no knowledge of motorsport', which was the primary reason for the agreement that he spend the year with Prodrive. The decision to partner came from the analysis that had been undertaken in the mid-1980s. Essentially, the analysis had been concentrated around the assumption that the company's output was too small to enable the company to survive in the industry as a manufacturer, competing on volume and price. The world's mass manufacturers were capable of producing at least three times Subaru's output and therefore, any attempt at price competition would lead to obliteration. It was the self-recognition that Subaru, as a small manufacturer could only survive 'with a distinctive, original production concept'. As Yamada expressed it 'Giant makers soon imitate and we lose originality and competition'.

It was therefore clear that the company had some very difficult choices if they were to make the strategy a reality. The company had produced an extremely good road car concept that had a number of unique design features that were very difficult to replicate. These characteristics were:

- Lightweight, compact, low centre of gravity
- A low vibration boxer engine with a wide power range
- A 4WD system that was lightweight and very well balanced
- Very rigid, lightweight chassis
- One engine production line able to deal with all market demand

Furthermore, the boxer engine was a unique concept that only one other manufacturer had developed and the Legacy provided the company with a basic car that was to prove suitable for developing as a superb rally vehicle. The point of this is that the car had not originally been designed with such a purpose in mind, although Yamada's fundamental belief in the value of engine power was a key element in the second part of the partnership equation.

The management of Subaru had already decided that there were a number of distinct advantages in a partnership, particularly with a specialist-engineering group who had motorsport experience and expertise. Firstly, there were a number of manufacturers who had successfully enhanced their brand through motorsport, including Vauxhall and Lotus, Ford and Cosworth, Mercedes Benz and AMG, Honda and Renault with Williams Grand Prix. Secondly, a partnership would provide the advantage of shared development costs for new exclusive models. Thirdly, companies such as AMG, Williams and TWR could reduce the design and development times for new market concepts. Fourthly, the added brand value of a link with such a company would provide distinctive brand advantage instantly. Finally, they would need to choose a partner who would add complementary skills, for example at motorsport, to their own skills of building very good road cars.

However, the experiences they had had with other partners in attempting to develop a Formula 1 engine had convinced both Mr Kuze and Mr Yamada that it was essential that there should be a mutual advantage in the partnership, beyond money. In an interview with Mr Koseki in December 1999 (in 1989 he was a manager of Vehicle Development as well as the leader of motorsport activities at FHI, and has now retired from Vehicle Development but still serves FHI as representative to the FIA and Japan Automobile Federation. He is a member of the Records Committee of FIA), he made a number of key statements that illustrate the company's philosophy towards the partnership.

Firstly, he was extremely clear that the relationships that had developed over the years between 1980 and 1989 when Subaru had participated in the Safari Rally had provided a number of advantages, including knowledge of

Prodrive's facility at Banbury, Oxfordshire, England.

Rothman's Porsche 959 from 1985, competing in the Pharoah's Rally

The Subaru Impreza World Rally car driving with Prodrive All Stars

Subaru's 2 liter 'Boxer' engine

Prodrive's customer rally team workshop

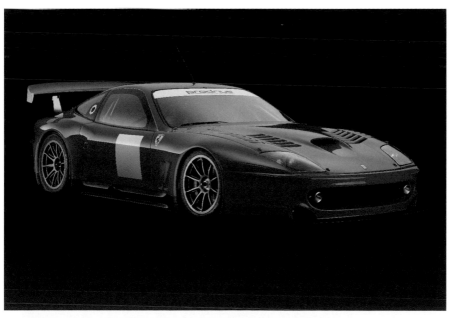

The Maranello prepared by Prodrive for the FIA GT series, 2001

Prodrive Subaru Impreza World Rally car, 2001

The Subaru Impreza P1 road car, 2001

Prodrive's new facility at Warwick

David Richards, Chairman and founder of Prodrive

Mr Tanaka, President and Chief Executive Officer, Fuji Heavy Industries

The Ford Mondeo BTCC Super Touring car 1999 and 2000

Colin McRae and Derek Ringer on their way to victory in the Legacy RS at the 1991 Talkland International Rally

Possum Bourne drives the Legacy RS in the Asia-Pacific championship

the sport and the participants. Secondly, the company had realized that they did not have the knowledge to compete independently in the WRC, at least not to win on a cost-effective basis. Thirdly, at the interview in 1999 he agreed that currently Subaru were the pupils in motorsport, but he did envisage a time when they would be in a position to undertake motorsport engineering and organization themselves. This latter point was very clearly made by the current President and Chief Executive Officer of FHI, Mr Takeshi Tanaka, in an interview with the author in December 1999.

4

Prodrive and Fuji Heavy Industries: The Development of the Relationship

Prodrive in 1989

Founded in 1984 by David Richards and Ian Parry, respectively Chairman and Commercial Director, the company has had a continuing record of profitability and increasing turnover. However, as the current Chairman has stated:

'In those formative years were a period when we tried our hand at a wide range of business initiatives. In hindsight, we were like so many new ventures, and often failed to focus on our core skills'.

The company certainly lost little time in rejecting that somewhat familiar fragmentation of effort, and swiftly developed a rarely found focus in the degree to which it recognizes its strengths and core skills, and applies them. In the early years of the company the concentration on the combination of marketing and engineering skills was a natural combination given the skills of the founding Directors. Both had had successful careers in marketing and, in David Richards's case, also in rallying and race preparation. In this context the mission statement is extremely clear: 'To provide a commercial and marketing platform for the successful promotion of clients'.

In 1984 Prodrive formed its first motorsport team, the Rothmans Porsche Rally Team and within its first year had won the Middle East Rally Championship and second place in the European Rally Championship. During that first year the team scored a total of seven rally wins, a clear indication of the success to come. In the 1985 season the Prodrive Rothmans Porsche Team retained the Middle East Rally Championship and won the Irish Rally Championship. The company's commercial success was also conspicuous and by the end of the second season of operations the company was relocated to its present headquarters at Banbury in Oxfordshire.

In 1987 Prodrive's successes in rally competition convinced the Bavarian motor manufacturer BMW that the company was the preferred partner to lead the company's World Rally Championship and touring car campaigns. In the first season of the World Rally Championships, Prodrive claimed its first

WRC win with the BMW Rally car, in the Rally of Corsica. In addition to the successful launch of the BMW WRC Team the company gained a total of five rally wins with both BMW and Porsche, including second places in the Middle East and French Rally Championships. On the track the Prodrive BMW M3 contested five rounds of the British Touring Car Championship.

In 1988, the first full year of competition, the Prodrive-prepared BMW M3 won the British Touring Car Championship and the marque's rally successes continued. The Prodrive BMW rally cars won the Belgian Rally Championship and second places in the French and Italian Championships, and in the process gained a total of thirteen rally wins for the company, a record. In the following year the BMW Prodrive M3 won their second British Touring Car Championship and the rally cars won the Belgian and French Rally Championships adding a further seven rally wins to the company's record. By 1989 Prodrive were a conspicuous and very successful company in the field of developing rally and track cars to enable manufacturers to capitalize on the marketing benefits of motorsport.

The company had proved its effectiveness in the rarefied atmosphere of World Rallying and the British Touring Car Championship. In addition the company's commercial stability and management expertise were continually developed as the company's founders determined to build a successful commercial and engineering team to ensure continued success. A decade later, many of the people who were employed at the start of the company remained part of a clearly outstanding organization. In the same way that Mr Yamada emphasized the vital importance of the 'soft management issues', the current Chairman of the company and founder Director David Richards has stated: 'I firmly believe that recognising the talents of an individual within a team environment is one of the greatest management challenges we continue to face.'

The company continues to build on the success it has achieved through the development of new skills and ideas, within the context of adding value to the customer's motorsport campaigns.

The Development of the Campaign: Partnering with Prodrive

The Beginnings 1989 to 1993

Once the decision to enter the World Rally Championship was taken, the managers most involved, Kuze, Narita and Yamada, confirmed that the most urgent question was, who to select as a partner? Relationships between Mr

Kuze, Mr Yamada and the Prodrive team, particularly David Richards, had developed over the following five years and late in 1989, the company approached Prodrive. At the time Prodrive's rallying activities with BMW were approaching a critical point as the Bavarian company had failed to achieve homologation of their own AWD saloon. Prodrive's Chief Engineer, David Lapworth was placed in charge of evaluating the new Legacy and the boxer engine as a potential WRC car. As Yamada said, the choice was not simply one way since both companies saw complementary strengths in the other.

David Lapworth's evaluation of the new Legacy, its AWD drive chain and new boxer engine was thorough and, as one would expect in the world of motorsport, very rapid. The Legacy had a number of advantages over its rivals including the stiffness and the relatively light weight of the chassis. The boxer engine gave a very low centre of gravity and the all wheel drive system, being symmetrical, offered a very low polar moment of inertia and therefore extreme controllability. The engine was not sufficiently powerful enough in its original form but Subaru were sure that they would be able to increase the power output considerably.

After the year spent with Prodrive and the extensive opportunity he had to observe the team in operation, it was Mr Yamada's strong recommendation that Subaru partner with Prodrive. This was not only based upon his clear perception that, as with many of the companies that were potential partners, Prodrive had the technical expertise to produce results. It was also based upon what both he, Mr Kuze and Mr Koseki all described as the vital, softer issues involved in a partnership.

The culture and attitudes of Japanese business are difficult to encapsulate, as with any other culture. However, there appear to be several unique characteristics in Japanese business culture, which, as far as Anglo-Saxon attitudes are concerned, are seemingly difficult to understand. When the Japanese are involved in business relationships, they are usually long term, collaborative, and characterized by the objective of achieving mutual advantage. (As a digression, it is for that reason the British government should not have been surprised by Honda's response to the Rover sale to BMW). There were a number of factors in the way Prodrive worked that impressed the key managers of Subaru, including the collaborative way the rally team went about solving problems. These factors were of considerable importance in the choice of Prodrive, as both Mr Koseki and Mr Yamada was insistent that these 'softer' issues should not be underrated both in terms of the choice of a partner but also in the development of the relationship.

The first factor that Yamada identified was, as he described it, 'the team's culture was a problem solving, not a blame culture. The Chief Engineer David

Lapworth is a person interested in finding solutions to win, not trying to blame losing on an individual'. This, as far as Mr Yamada was concerned, mirrored the fundamental concept that governs the team's efforts. As he expressed it 'rationality for victory is the basis for co-operation'. Furthermore, after his experiences of the abortive attempts with the Formula 1 engine and chassis he found that the procedures at Prodrive were favourably open. At least as far as he was concerned, the need for trust and for a clear understanding of the role and contribution of each of the cultures was paramount. But the concepts of how the two cultures were to work toward their common goals were the subject of considerable thought.

The soft aspects required that the joint partnership evolved into the 'creation of a new culture' where all standards are based on 'the rationality to win'. Within this cultural aspect of the partnership Yamada described two parts of 'the strategy in culture' which the two companies developed. The first was the concept of recognizing and respecting the other cultures involved, and the second was the recognition that if they were to create a 'new culture' they had to identify the elements. He described Prodrive's culture as being essentially 'a hunting peoples culture' whereas in his perception the FHI and STI companies have 'an agricultural peoples culture'.

The 'concrete method' that Yamada defined was a very practical and clearly defined set of activities, which were to prove extremely far sighted. The allocation of responsibilities between the two partners does not appear to have altered since the beginning and were, according to Mr Yamada, based upon his clear vision of what each partner was 'best at'. In Yamada's view it was essential that Prodrive concentrated on developing the rally car, using the road cars that Subaru were very good at building; 'The work sharing was based upon the strong points of each partner and on advantageous location'. Essentially these two principles meant that Subaru Tecnica International supplied the engine, electronics and electrical controls and Prodrive developed the modified body, chassis, tyres and transmission. In addition Prodrive undertook to manage and operate all the necessary testing and rally operations, including driver selections.

SUBARU/PRODRIVE TASK BREAKDOWN
Subaru: Engine and electronics, electric control
Prodrive: Body, chassis, tyres, transmission

These concepts formed the basis of the original partnership agreement, which was struck between the two companies on 20 September 1989. In this first year the total budget was in the region of £9 m.

Shortly after the intensive development programme to transform the Legacy road car into a rally winning race car was started in the early months of 1989, it was apparent that the boxer engine was not producing sufficient power. The initial set of engine modifications was undertaken by STI, but early progress was slow and the results disappointing. After several months Prodrive were becoming somewhat exasperated, as David Lapworth knew only too well that power was crucial; as did Mr Yamada. The problem was how to increase the output of the engine without sacrificing reliability or torque. The issue was that the engine was at the heart of the whole development programme and until the power issue was resolved, other factors were put on hold.

On 21 August 1989 David Richards received a request from Mr Yamada concerning the analysis Prodrive's engineers had undertaken of the Boxer engine: 'Regarding the Boxer – what are your recommendations for engine modifications?' It had become apparent, that if the power output from the Boxer engine was to be increased to provide a performance envelope capable of winning rallies before the start of the 1991 season, urgent action was needed. Over the course of the winter a number of design changes were made to the Boxer engine's inlet manifold and port design, much of the work being completed by the Prodrive engineering team. The request from Mr Yamada was fulfilled and the required power enhancements achieved.

Prodrive continued to run the BMW M3 team during the 1989 and 1990 seasons, recording a total of eleven victories in international rallies; increasing the company's total of victories to forty-six. The company's engineers were working to develop the chassis, body and transmission for the first assault on the World Rally Championship in 1991. Over the period from June 1989 through to the end of 1990, the relationship between the two companies was developing and the core of the relationship was at two levels. Mr Yamada and David Lapworth dealt with the technical matters and David Richards and Mr Kuze concerned themselves with the contractual relationship, although it is never quite that simple.

The team began the search for a major component of success, the drivers and co-drivers, early in 1990 and during the following year utilized the services of a number of drivers in pursuit of the perfect combination of man and machine. In those early days the budget for the season (as noted above) was in the region of £9 m. This figure was to cover the development of the rally cars, the team support and transportation and everything needed for the team's participation in the World Rally Championship. Although Prodrive was committed to obtaining sponsorship for the team's activities, this had proved less than easy to obtain. The team continued to run the Legacy cars in the WRC and it was not until the 1994 season and with the new Impreza, that the team acquired the BAR '555' sponsorship which doubled the available budget.

The initial development work on the first Legacy rally car had been completed in time for the 1991 season. At least the first stage had been achieved and the company had a car with which to compete. During the first season of competition, which was accepted by FHI as a development period, the team scored a number of international and national victories with different driver/ co-driver teams. Notable among them was the name of Colin McRae.

Table 4.1 **1991–93 Rally Results**

1991

Event	Country	Driver/Co-Driver	Car	Wins
Talkland Int	UK	C McRae/D Ringer	Subaru Legacy RS	47
Circuit of Ireland	UK	C McRae/D Ringer	Subaru Legacy RS	48
Scottish Int	UK	C McRae/D Ringer	Subaru Legacy RS	49
Manx Int	UK	C McRae/D Ringer	Subaru Legacy RS	50
1992				
Vauxhall	UK	C McRae/D Ringer	Subaru Legacy RS	51
Pirelli	UK	C McRae/D Ringer	Subaru Legacy RS	52
Scottish	UK	C McRae/D Ringer	Subaru Legacy RS	53
Ulster	UK	C McRae/D Ringer	Subaru Legacy RS	54
Manx Int	UK	C McRae/D Ringer	Subaru Legacy RS	55
Elonex	UK	C McRae/D Ringer	Subaru Legacy RS	56
1993				
Vauxhall	UK	R Burns/R Reid	Subaru Legacy RS	57
Pirelli	UK	R Burns/R Reid	Subaru Legacy RS	58
Scottish	UK	R Burns/R Reid	Subaru Legacy RS	59
Indonesian	Indonesia	P Bourne/R Freeth	Subaru Legacy RS	60
New Zealand	NZ	C McRae/D Ringer	Subaru Legacy RS	61
Malaysia	Malaysia	C McRae/D Ringer	Subaru Legacy RS	62
Manx Int	UK	R Burns/R Reid	Subaru Legacy RS	63
HK-Beijing	China	A Vatanen/B Berglund	Subaru Legacy RS	64
Thailand	Thailand	P Bourne/R Freeth	Subaru Legacy RS	65

But success in international motorsports is never easy and the foundations of the philosophy that Mr Yamada had outlined in 1989, was one of the core reasons for the success of the initial development period. Another factor was the clear and unambiguous separation of responsibilities within a framework of flexible co-operation that always recognized that: 'Rationality for Victory the Base of Co-operation'.

The team improved the car's performance continuously during the period 1990 to 1993, as well as helping Subaru with the development of the new Impreza, which they launched worldwide in 1992. The team was not only building the Legacy's performance and reliability but also developing a close working relationship with the motorsport arm of Fuji Heavy Industries, Subaru Tecnica International or STI. The Subaru Legacy had always been regarded as an interim solution on which the two companies could build their relationship whilst development of the Impreza continued.

The Subaru Legacy RS claimed several national rally wins from 1991, and in 1993, the team's first full season, they won their first WRC rally, in New Zealand. This was the first year of the company's involvement in the World Rally Championship and the Legacy had already made its impact on the motorsport world. The Subaru team had taken places in the Acropolis Rally in Greece, the Safari Rally, the Thousand Lakes Rally in Finland and rallies in the United States, Australia and New Zealand. These activities had helped to bring the new car to the attention of a wider motoring public and achieved greater market exposure than would have been possible by other means. These results proved to Fuji Heavy Industries/Subaru that their initial strategy in choices of partner, Prodrive, and the arena, the World Rally Championship, had been correct. The company had used the image of rallying success and the reliability and performance of their cars to demonstrate to consumers the quality of their product.

Whilst 1993 was the first full year of the team's participation in the World Rally Championship, it was also the last year that the team used the Subaru Legacy as the basis for the rally car. Since Subaru had launched the Impreza in 1992, David Lapworth and the Prodrive engineers had recognized that the chassis layout and design were an excellent platform for a new rally car. Development work had begun in 1992 and the new car was available for the start of the 1994 World Rally Championship. Results in the first year were impressive and the cars, driven by Carlos Sainz, Colin McRae and Possun Bourne collected six wins in the season. Development of the car continued throughout the season with performance improving all the time. The team was certainly delivering the results that the Subaru marketing team had hoped for. During several extended interviews with senior members of Fuji Heavy Industries in 1999, it was stated several times that 'The team's performance has always exceeded our expectations'. (Mr Yamada, President of Subaru Tecnica International, December 1999).

The results of the partnership speak for themselves. In the Subaru team's first foray into international rallying events, the Subaru Legacy RS driven by Colin McRae and David Ringer won the Talkland International Rally. The same pairing of car and driver completed three more wins in that first season.

As the team's experience of the car developed and the relationship and technical co-operation between STI and Prodrive's engineers grew, results improved in each successive season. In the following two years the Subaru Legacy's team collected fifteen international rally victories, six in 1992 and nine in 1993.

The impact of the success in the World Rally Championships upon the market place was substantial and the marketing efforts of the manufacturer and dealers worldwide placed a great deal of emphasis on the programme. In 1993 Subaru's sales worldwide amounted to 511 628 units, but of more than 310 000 sold in Japan, more than 200 000 were mini vehicles. Therefore, the company determined to increase their marketing effort through motorsport within the Pacific region. The company utilized the expertise of Prodrive to support an assault on the Asia-Pacific Rally Championship using the new Subaru Impreza 555 and from 1993 to the final round of the Championship in 1997, their record of success was unprecedented.

Asia-Pacific Rally Championship

1993 Drivers' and Manufacturers' Champions

1994 Drivers' and Manufacturers' Champions

1995 2nd Drivers' and 3rd Manufacturers' Championship

1996 1st Drivers' and 2nd Manufacturers' Championship

1997 Drivers' and Manufacturers' Champions

Maintaining Momentum and Generating Competitive Advantage

Building on Success: The Period 1994 to 1999

The new season was the culmination of an extensive technical development programme that had begun in 1992, before the new Subaru Impreza had been launched worldwide. The car's success in the 1994 season was to prove only an indication of its potential. The 1991–93 seasons had proved conclusively that the team had managed the transition from one manufacturer (BMW) to a new one (Fuji Heavy Industries/Subaru) and moreover, that the intensive four year development programme had been successful. The very complex business of establishing the necessary relationships between two very different entities at all levels is a crucial factor, indeed a necessary condition, for successful participation in World Class motorsport.

In the 1993 WRC season the team gained the sponsorship of the British American Tobacco Company and ran the Legacy with the blue and yellow livery and the '555' logo. In 1994 the new Subaru Impreza 555 also entered under the sponsorship of the 555 tobacco brand, which had doubled the available budget from £12 m to £22 m in 1994, with Fuji Heavy Industries supplying £10 m of the total.

The fact that Prodrive and Subaru Tecnica International had very clear guidelines as to the different roles and contributions each would bring to the effort, was a vital ingredient in their early success. In interviews with Mr Kuze, President of STI in the early years of the campaign and Mr Yamada, then in charge of the engine development and the current President of the company, both mentioned two essential ingredients of a successful sponsorship relationship. The first is the absolute trust that is necessary between the parties, a factor that it is difficult to overemphasize, as the sums of money involved are very considerable. The second, is the clear and unambiguous recognition by each party of their respective contributions and responsibilities towards the success of the effort. STI were totally clear from the beginning that they would leave the development of the rally car to Prodrive; that was Prodrive's capability. STI would concentrate upon applying what they could learn from technical developments of the rally cars and on developing better road cars.

In spite of constant improvement, successes had eluded the team with the Subaru Legacy. The new Subaru Impreza 555 was to prove a vastly more competitive car. In 1994, the first year of competition with the new car, the team recorded three World Rally Championship and three Asia Pacific Rally Championship wins, and secured second place in both the WRC Manufacturers' and the Drivers' Championships. The following year, in the 1995 Championship, the Subaru team achieved their first domination of the event, winning the Manufacturers' and Drivers' Championships with the new Subaru Impreza 555. They followed this victory with further Manufacturers' Championships in 1996 and 1997, taking second place in the Drivers' Championship in each of those years.

Throughout the rest of the decade the Subaru World Rally Team continued to be a major competitive force in the World Rally Championship and although the successes of the 1995–97 seasons were not repeated, FHI's financial and commercial success continued. The consistency with which the team produced effective results that kept the Brand and the company's cars in the international rallying limelight were an essential ingredient in the sustained sales and marketing effort.

Table 4.2 **1994–96 Rally Results**

Event	Country	Driver/Co-Driver	Car	Wins
1994				
Acropolis	Greece	C Saintz/L Moya	Subaru Impreza 555	66
New Zealand	NZ	C McRae/D Ringer	Subaru Impreza 555	67
Malaysia	Malaysia	P Bourne/T Sircombe	Subaru Impreza 555	68
Australia	Australia	C McRae/D Ringer	Subaru Impreza 555	69
HK-Beijing	China	P Bourne/T Sircombe	Subaru Impreza 555	70
RAC	UK	C McRae/D Ringer	Subaru Impreza 555	71
1995				
Monte Carlo	France	C Saintz/L Moya	Subaru Impreza 555	72
Portugal	Portugal	C Saintz/L Moya	Subaru Impreza 555	73
Indonesia	Indonesia	C McRae/D Ringer	Subaru Impreza 555	74
New Zealand	NZ	C McRae/D Ringer	Subaru Impreza 555	75
Catalunya	Spain	C Saintz/L Moya	Subaru Impreza 555	76
RAC	UK	C McRae/D Ringer	Subaru Impreza 555	77
1996				
Thailand	Thailand	C McRae/D Ringer	Subaru Impreza 555	78
Acropolis	Greece	C McRae/D Ringer	Subaru Impreza 555	79
Malaysia	Malaysia	K Eriksson/S Parmander	Subaru Impreza 555	80
Sanremo	Italy	C McRae/D Ringer	Subaru Impreza 555	81
Catalunya	Spain	C McRae/D Ringer	Subaru Impreza 555	82

Table 4.3 **1997–99 Rally Results**

Event	Country	Driver/Co-Driver	Car	Wins
1997				
Monte Carlo	France	P Liatti/F Pons	Subaru Impreza 555 WRC97	83
Sweden	Sweden	K Eriksson/S Parmander	Subaru Impreza 555 WRC97	84
Safari	Kenya	C McRae/N Grist	Subaru Impreza 555 WRC97	85
Corsica	France	C McRae/N Grist	Subaru Impreza 555 WRC97	86
China (A-P)	China	C McRae/N Grist	Subaru Impreza 555 WRC97	87
NZ	NZ	K Eriksson/S Parmander	Subaru Impreza 555 WRC97	88
San Remo	Italy	C McRae/N Grist	Subaru Impreza 555 WRC97	89
Australia	Australia	C McRae/N Grist	Subaru Impreza 555 WRC97	90
RAC	UK	C McRae/N Grist	Subaru Impreza 555 WRC97	91

Event	Country	Driver/Co-Driver	Car	Wins
1998				
Portugal	Portugal	C McRae/N Grist	Subaru Impreza 555 WRC98	92
Corsica	France	C McRae/N Grist	Subaru Impreza 555 WRC98	93
Acropolis	Greece	C McRae/N Grist	Subaru Impreza 555 WRC98	94
555 China	China C	McRae/N Grist	Subaru Impreza 555 WRC98	95
1999				
Argentina	Argentina	J Kankkunen/J Repo	Subaru Impreza 555 WRC99	96
Acropolis	Greece	R Burns/R Reid	Subaru Impreza 555 WRC99	97
1000 Lakes	Finland	J Kankkunen/J Repo	Subaru Impreza 555 WRC99	98
Australia	Australia	R Burns/R Reid	Subaru Impreza 555 WRC99	99
Great Britain	UK	R Burns/R Reid	Subaru Impreza 555 WRC99	100

The competition results speak for themselves and the highly successful partnership has enabled a small, niche manufacturer to compete and win successive world championships in one of the most competitive and difficult arenas of motorsport in the world. However, Subaru is a business and the ultimate goals were related to profits and sales success; winning championships and improving the brand were the means to do so.

---

(5)

The Results of the Relationship

The Development of Subaru's Market Strategies from 1993

The logic behind Subaru's foray into the World Rally Championship was formulated and driven by the fact that, as a niche player in an intensely competitive global industry, they had to find a differentiating factor. Poor profit and sales performance in the mid-1980s (see page 58) had convinced the Board and management of the company that in order to stay an independent player the company had to 'raise its image and build on its non-replicable strengths'. Furthermore, whilst at the end of the 1980s the company had a well established reputation as a manufacturer of all-wheel-drive vehicles their model sales profile was low volume low profit. Therefore, their decision to pursue a policy of image and brand building through motorsport to alter that profile was formulated on a clear set of strategic goals. That is not in itself particularly innovative or new, but what was unique amongst the Japanese manufacturers in the 1990s, was the clarity and persistence with which they formulated their plans.

The period 1990 to 1995 served as the basis for Subaru to consolidate the clear set of marketing strategies that had been formulated during the last years of the 1980s. As stated above, there were three related technical, product based facts, upon which Mr Kuze (then Chairman of Subaru Tecnica International) and the Board of Fuji Heavy Industries led by their Chairman and President Toshihiro Tajima, based their new strategies. These were firstly, the company's long history in the development of all wheel drive technology for relatively small, low powered cars. Secondly, the new and successful Boxer engine which, in conjunction with the company's expertise in turbocharging provided a potentially powerful unit with a low centre of gravity. Thirdly, the new Legacy that had been launched and on which the company had (along with the Boxer engine) 'bet the company's future'.

The company had a clear set of marketing objectives that were stated in 1989 and have not changed. These principles were to:

• Provide a platform for the company to place the characteristics of the products before current and potential customers.

- Provide an additional opportunity and means for the technical development of the products.
- Provide a platform for the development of the brand and image of Subaru, both domestically and overseas.

The company has sustained their strategy throughout the decade and remained focused on their primary objectives of moving the product range from its previous dependence on mini cars and vans and differentiating the product range in unique ways. In a presentation that was written in 1989 Mr Yamada, then the company's head of motorsport under Mr Kuze the President of STI, emphasized that as a small manufacturer Subaru had to recognize that the company needed to accept that:

SELF-RECOGNITION
SUBARU IS A SMALL MANUFACTURER
Small Makers Could Survive with Distinctive, Original Production
Because the Industry is a Rat Race!

This foundation is now recognized by many manufacturers and is indeed a major element in the niche marketing policy of many producers in overcrowded markets who are unable or unwilling to continue to compete on price alone. But the analysis that they had undertaken went rather further than that. The company also defined the key elements of their marketing and product offering that was to be encapsulated within what was defined as the 3S concept. This was defined as:

SPEED
STYLE
SPACE

However, the most important and unique aspect of the campaign was that the marketing strategy was built around the technical reality of the new Legacy, the platform and its engine and transmission layout. Therefore, there was a fundamental synergy within the anatomy of the brand's objectives and the attributes of the product. Theory says that there are three elements to the anatomy of a brand (see Figure 5.1).

It is important to recognize that that is exactly what Subaru achieved and it is an essential factor in the success of their campaign that the product's characteristics were part of the facts behind the marketing campaign. The campaign mirrored and indeed emphasized the facts of product design and performance and was aimed at precisely those consumers who would appreciate those characteristics.

The other vital ingredient of such a campaign that is a recognized factor for a small manufacturer, is that it must be difficult to replicate. This was also

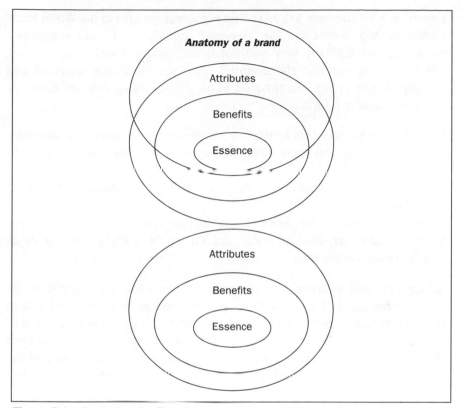

Figure 5.1 Anatomy of a Brand

evident in the company's original analysis both of their potential product offering and their marketing. The difficulty for any small manufacturer is that the larger players are always attempting to enter niche markets and many (such as with the AWD Range Rover, APVs or SUVs) will try to create niches. If they see a small player pursuing a niche they will attempt to copy it, thereby destroying the originality of the initial concept and destroying the original players' competitive advantage. Subaru were rightly confident that their products' technical characteristics, the combination of the boxer engine and turbocharging, all wheel drive and specific weight and dimension advantages, made it extremely difficult to imitate.

All of these factors were recognized by management and the marketing team and the company's campaign was clearly focused on emphasizing those characteristics and building upon them.

The company's marketing efforts are organized into domestic and overseas sales divisions and all of their marketing efforts are driven by that split. The policy was to develop their marketing campaigns around the launch of the

Legacy in 1989 and then to add to that the developments in the World Rally Championship campaign. The essential ingredients of the company's marketing policies were very traditional and lacked a certain innovation, when you consider that the essential logic was so clearly analysed and developed. The policy concentrated upon the following primary elements that are similar in both areas:

1. Worldwide promotional activities co-ordinating with national dealerships.
2. Advertising and brochure campaigns emphasizing the participation and success of the team in the WRC.
3. Media coverage in all areas (press, TV and so on) involving the WRC campaign.
4. Specific dealership promotions using the rally cars.
5. Promotional activities at WRC and Group N Rally events involving dealers and distributors.

Domestic and overseas marketing departments are convinced that the WRC campaign had enhanced the product image, and had undertaken several consumer surveys that showed a distinct change in the perception of the Subaru brand. During the decade of the programme these surveys have shown a gradual increase (but varied between markets) in awareness of the product's qualities, particularly in the major European markets of Germany, Italy and the United Kingdom. These surveys demonstrated that the attributes of Subaru products embodied in the Rally and Road cars, namely all wheel drive and performance, were critical to the Brand Image, and firmly established in terms of the following characteristics by the end of the 1990s, after nearly ten years of the campaign.

• Superior AWD technology
• Good performance in motorsports
• Fun to drive
• High power, high performance
• Uniqueness, originality

However, in surveys carried out by the departments in May/July of 1996 and in May of 1998 the results were somewhat illuminating. In Germany, where the Impreza Turbo was not sold and rally success not promoted, the image of the car did not alter much. The product image as a performance motor car, with unique engine design and having advanced AWD technology, actually became weaker amongst both car owners and Subaru owners. In Australia the same characteristics improved and in particular the Subaru users related the brand to advanced AWD technology.

The most recent survey conducted in the same markets (Italy, Germany and the United Kingdom) in March 1999, confirmed the continued improvements in the product's brand perception, by both Subaru and other car owners. In every characteristic the brand image was affirmed in a larger percentage of interviewees. The programme was obviously successful in re-enforcing the brand's characteristics and defining the brand in a manner that was compatible with the objectives of the campaign. The major conclusion from this aspect of the work undertaken by the company is that redefining a brand is not a quick or easy task. In this case the World Rally Championship campaign had lasted for nearly a decade and had achieved considerable success, including three WRC Manufacturers' Titles in 1995, 1996, 1997, one Drivers' Title in 1995 and two second places in the Drivers' Championship in following years.

There is therefore, strong *prima facie* evidence that the campaign had produced an impact upon the European market and that the campaign had had a similar impact within the domestic market. However, the development of the market strategies throughout Europe remained totally in the hands of Fuji Heavy Industries and their Subaru marketing divisions. In an extensive interview with Mr Yamada President of STI in December 1999 he emphasized that the division of responsibilities between Prodrive and Subaru was an essential ingredient to the success of the WRC campaign. David Richards of Prodrive confirmed the separation of responsibilities but also accepted that there 'were missed opportunities in the lack of co-ordination of the sales and marketing campaigns across Europe'.

David Richards also expressed the view that STI appeared to 'pigeon hole' Prodrive and that there were currently insufficient links between the two companies to ensure effective co-ordination of the marketing effort. He went on to express the view that initially Subaru were probably not prepared to believe that Prodrive would be able to beat Ford or Toyota in the early years, and so STI left Prodrive to get on with the rallying. This was certainly confirmed in interviews with Mr Kuze, the President of STI in 1991 and the current President Mr Yamada. Since the team won their first World Rally Manufacturers' and Drivers' Championships in 1995 and 1996, there have been closer ties between the two companies, particularly in the technical development and design of the road vehicles. David Richards again in an interview in May 2000: 'The launch of the new car in 2000 is an ideal time to develop closer marketing co-operation.'

Both companies accept that there is much to be gained from further co-operation in marketing and merchandizing all aspects of the campaign. The President of Fuji Heavy Industries Mr Takeshi Tanaka was clear that the marketing and merchandizing efforts were capable of further improvement.

Finally, the leading individual in charge of the company's motorsport campaign Mr Yamada, the President of STI, was unequivocal in his criteria for the success of the campaign: 'There is the vital importance of trust and that the value of the rally success is perceived to be earned and, of more value than the cost.'

Technology, Transfer and Road Car Applications

During the course of the initial negotiations STI submitted a Legacy saloon for initial technical evaluations, which were undertaken by Prodrive's engineering team, led by David Lapworth. Their findings, which were reached relatively quickly, indicated that the Legacy had a number of specific advantages as a rally car as a result of the basic layout and engine configuration. As Mr Yamada had stated 'There are unique advantages with the Boxer engine and the transmission layout'. However, it is important to remember that most of the technology on the Subaru cars in 1989 had not been derived from any motorsport activity. The All Wheel Drive system was developed for military vehicles and the boxer engine had been developed in-house in the early 1970s. The new engine was a development of the original concept.

More specifically, the advantages of the Legacy layout can be summarized as follows:

- The Boxer engine layout made it possible for the front stress members to be directly under the engine, offering a low centre of gravity.
- The Boxer engine is narrower than other engines and enables the suspension to have longer stroke, greater vertical movement and a lower centre of gravity.
- The engine unit offers a lower level of vibration and is more compact, than other units.
- The body shell is light in weight and very rigid.
- Finally, the engine is very difficult to replicate and proved to offer very flexible power delivery.

However, the initial trials of the Rally car in 1990 showed that the power output objectives had not been achieved and the car's performance was definitely suffering through a lack of power. In view of the fact that Mr Yamada had consistently stated that, in his view, a racing car with poor handling might be redeemed by higher engine power, the opposite is rarely true; this was a difficult problem for STI. Under the initial terms of the contract, the breakdown of responsibilities between Prodrive and STI had been very specific:

SUBARU/PRODRIVE TASK BREAKDOWN
Subaru: Engine and Electronics, Electric Control
Prodrive: Body, Chassis, Tyres, Transmission

Both Mr Yamada and David Lapworth emphasized that after the initial development period it became clear that the engine power output was inadequate for rallying and that without further engine development the car's performance would suffer. The result was that the two companies agreed to collaborate over the development of the Boxer engine. This led to a rapid and massive increase in the power output of the engine with the results that proved so successful in the cars' first year of competition. (See Table 3.3).

This relaxation of the initial delineation of responsibilities between the two companies proved to be an important milestone in the development of their close relationship. Whilst it had been agreed that the clear definition of what each company would undertake was a vital part in developing a smooth and efficient relationship, flexibility soon proved to be vital. As Mr Yamada made it clear in an interview in December 1999 the foundation of the agreement from the beginning was trust. He emphasized that the technical advantages of the basic design and layout of the Legacy and the technical abilities of Prodrive in race and rally technology, were both crucial and necessary elements for success. However, in his view at least as important were the 'soft' management issues that governed the manner in which the two organizations worked together. To quote him, the two companies:

- Have a high degree of trust in each other's ability.
- Rely on each other and do not interfere. Collaboration is not interference.
- The technological collaboration has meant that the car has improved both in terms of practical performance and the customer's perception of the car's qualities.

He then made a key statement which is probably a vital element in the sound level of co-operation that has continued to occur within the relationship and that has led to the continued improvements in the car's performance, on and off the rally circuit: 'The fact is that Subaru's performance objectives have always been exceeded as they have achieved more rally success earlier than they had planned'.

These attributes of the evolving relationship appear to have been laid down very early in 1990, as the Legacy was being developed from a road car to a successful rally competitor. The relationship was further cemented as Subaru started to finalize the evolution of their new car, the now famous Impreza. In this case the two companies collaborated from a very early stage, exchanging ideas about the external design of the road car so as to enhance the similarity between rally and road car and to optimize the stiffness of the chassis and body shell. The development programme was carried forward jointly in order to optimize the joint contribution to the effective performance of the rally car and the marketing effectiveness of the road car.

In this area of collaboration it is important to remember that the technology of transmission, drive chain, braking controls and the engine power of a modern rally car are totally unsuitable for normal road use. The Subaru Impreza 555 rally car is a very different automobile under the skin than the Impreza Turbo road car. Nevertheless, the vital connection that is made between the road car and the rally car's success is crucial to the commercial success of the marque. The fact is that the rallying activity enables the company to raise its image, differentiate its products and emphasize the importance of the differences in design through demonstrating the influence of design on performance.

Inevitably the transfer of technological developments for the rally car are going to have limited application to the road car. However, the continuous improvement in the technology that drives the rally car's performance does provide an additional platform for the further development of the technology of the road cars. Furthermore, as with the rapid improvements in engine performance that were achieved through joint collaboration in the early 1990s, the main advantage to the manufacturer is the extremely short time frame that such developments take when driven by the demands of international competition. The World Rally Championship is, in that respect, similar to Formula 1, 'War without bullets'. It is therefore a rather appropriate arena in view of the origins of Subaru's all wheel drive technology, a technological area in which the company has the longest experience of any automobile manufacturer in the world.

In this context, the fact that Subaru was the first Japanese manufacturer to install a Traction Control System in its road vehicles was directly related to the development of the transmission systems for the rally cars. The system uses both engine and brake control. This innovation simply confirms Subaru's position as one of the most innovative Japanese manufacturers, in spite of being the smallest.

The external design of the Legacy and Impreza were both influenced by the demands of the rally cars. In particular the outline of the Impreza wheel arches and other features of the bonnet design have been altered from the original concept as the demands of the rally car were satisfied. The end product is perhaps more aggressive, spectacular and a better expression of the speed and style characteristics so critical to the brand.

Finally, the recent development of special models by Subaru, Subaru Tecnica International Inc., and latterly Prodrive has provided the company with further opportunities to build upon the rally success. These cars are perhaps most closely technically related to the rally cars in both technical design and external appearance. The unique specifications and rally related history have provided the company with an opportunity to delineate a further market niche for their products as well as demonstrating to consumers the race-bred heritage of the technology and links between rallying, technology and driving performance in all its aspects. In the last two years of the 1990s demand for the Impreza models has continually exceeded supply.

Financial and Sales Results of the Campaign

In the company's 1990 Annual Report, the President and Chairman of the company, Toshihiro Tajima, reported that 'the sales of the Legacy had risen from 46 724 in 1988 (the year of its launch) to 63 793 by 1991'. In addition, he also reported that 'the Legacy series continued to expand its reputation in world markets' and that 'Sales of the Automotive Division increased 16.7 per cent', thereby generating '81.0 per cent of the Company's net sales'. This was in spite of the continued recession in the United States and Canada and, although there was an overall decline in total passenger car sales in the European market, Subaru's unit sales rose by 13.7 per cent.

This was the first year of the company's involvement in the World Rally Championship and the Legacy Model had already made its impact on the motorsport world. The Subaru World Rally Team had taken places in the Acropolis Rally in Greece, the Safari Rally, the Thousand Lakes Rally in Finland and rallies in the United States, Australia and New Zealand. These activities had helped to bring the new car to the attention of a wider motoring public and achieved greater market exposure than would have been possible by other means. In these early days the annual budget was in the region of £9 m and was funded almost entirely by Fuji Heavy Industries.

The company was able to progressively increase the sales of its Legacy model against the trend in both the domestic and overseas markets. In particular the company was replacing sales of the low value added mini vehicles with the premium priced larger vehicles such as the Legacy (1989–94) and the Impreza from 1992.

The key statistic is that in all markets the Legacy and Impreza contributed a greater proportion of total sales. In the domestic market in 1989 out of total sales of 312 992 vehicles, 225 062 were of mini passenger cars and vans; that is 70 per cent. Another important fact is that in the company's domestic market, whilst national sales fell for the third time in three years (falling 7 per cent in 1992) Subaru's sales rose by more than 1 per cent. The fall in sales for other domestic manufacturers ranged from Isuzu 21.1 per cent, 18.2 per cent for Nissan to 7.7 per cent for Toyota. In a fax to David Richards the Chairman of

Table 5.1   **Sales Figures for 1988–93 Domestic and Overseas**

Domestic Market

	1988	1989	1990	1991	1992	1993
Mini passenger van	143 672	125 805	126 284	115 399	118 328	112 502
Mini truck/van	114 862	99 257	126 023	123 202	104 763	96 799
'Legacy'	–	46 724	57 542	63 793	62 668	62 125
'Impreza'	–	–	–	–	7 118	31 319
'Forester'	–	–	–	–	–	–
Others	69 458	41 092	27 436	20 407	14 341	7 773
Total	**327 992**	**312 878**	**337 285**	**322 801**	**307 218**	**310 518**

Overseas Market (Export + Local production)

	1988	1989	1990	1991	1992	1993
Mini passenger van	9 071	6 826	6 778	7 899	8 402	6 239
Mini truck/van	–	–	–	–	–	–
'Legacy'	–	98 360	110 432	126 534	122 978	76 244
'Impreza'	–	–	–	–	1 790	73 915
'Forester'	–	–	–	–	–	–
Others	263 137	125 797	119 515	140 173	115 161	44 712
Total	**272 208**	**230 983**	**236 725**	**274 606**	**248 331**	**201 110**

Prodrive, Fuji Heavy Industry's Head of STI Mr Kuze stated:

> 'It proves that high evaluation on Subaru as cars for drivers has begun to be spread over the Japanese market. I appreciate that SWRT's brilliant activities in WRC and APRC, to which Prodrive is a main contributor, and Championship in the '93 All Japan Rally Championship are the two big factors in bringing about such an evaluation.' (Kuze 13 January 1994).

During that period the company's sales of the mini car and truck range fell and sales of their new vehicles rose, improving profitability and turnover. Mini vehicle sales fell from a total of 209 310 units (67 per cent of total sales) in 1993, to 163 782 units (53 per cent of total sales) in 1997, whilst sales of the Legacy, Impreza, Forester and other models grew from 101 217 units to 140 707 units. As importantly, the company's total exports of vehicles (which are exports plus local production) remained at approximately 40 per cent of total sales from 1990 to 1993. The company's turnover from vehicle sales in the period from 1991 to 1993 rose from 611 893 m (overseas sales ¥238 520 m), 81 per cent of total sales, to ¥723 711 m (82.9 per cent of total sales) in spite of severe economic difficulties in the overseas sales markets, where unit sales actually fell (see Table 5.2).

Table 5.2 **Company and Financial Details from 1991 to 1999**

Auto Sales (¥/m)		Gross Profits (¥/m)	
1991	674 684	1991	71 333
1992	723 711	1992	103 373
1993	618 816	1993	117 984
1994	603 000	1994	84 402
1995	687 000	1995	120 612
1996	602 000	1996	133 590
1997	1 067 199	1997	178 260
1998	1 146 215	1998	276 207
1999	1 198 173	1999	284 652

The company's rally successes and related marketing efforts, at least from raw sales statistics, appear to have had an impressive impact upon both domestic and overseas sales. The company's sales revenues have moved from being reliant on low value added mini vehicles and their domestic market to a sales profile dominated by higher value added vehicles and overseas sales. In 1992, sales of the company's main models, the Legacy, Impreza and Forester, rose from a total of 83 584 (domestic) and 239 929 (overseas) units, which was 58 per cent of total sales. By 1999 these figures had increased to 124 341 (domestic) and 271 218 (overseas) units, which was by then 69 per cent of total sales. There is little doubt that some of the increase in overseas sales was due to the decline of the Yen between 1997 and 1998. However, the trend has continued into 1999 in spite of the Yen's rise against the Dollar; from July 1999 to July 2000 the Yen rose from ¥122/$ to ¥107/$.

By the end of the twentieth century Fuji Heavy Industries' financial position was stronger than some of the much larger Japanese automobile companies (see Economist Intelligence Unit Report, Japanese Motor Business, 3rd Quarter 1998). Their financial position has benefited from strong sales in the United States and Europe, which has reduced the company's reliance on its domestic market. The strong sales performance of the Forester in the United States and Europe supports the company's view that rally performance has provided a strong brand image that is perceived to be relevant to all the company's products, not just the Impreza.

The sales figures from 1996 (see Table 5.3) represent a fundamental shift in the company's product profile and sales profile and, as a result, potential profitability. Continued improvement in the product, the increasing proportion of overseas sales, and the more fundamental shift from the low value added mini cars/vans to the higher value added niche vehicles, are the

Table 5.3   **Sales Results 1994–99**

Domestic Market

	1994	1995	1996	1997	1998	1999
Mini passenger van	122 684	115 867	130 882	89 410	88 330	107 542
Mini truck/van	94 884	88 025	80 496	74 372	62 079	70 975
'Legacy'	88 924	88 490	91 478	66 576	59 686	70 181
'Impreza'	35 916	44 986	40 722	38 986	36 844	29 934
'Forester'	–	–	–	33 388	29 108	23 010
Others	8 506	7 374	4 913	2 751	1 923	422
Total	350 914	344 742	348 491	305 483	277 970	302 064

Overseas Market (Export + Local Production)

	1994	1995	1996	1997	1998	1999
Mini passenger van	6 534	5 522	4 116	6 216	3 628	2 973
Mini truck/van	–	–	–	–	–	–
'Legacy'	89 106	103 982	125 281	129 200	131 504	129 494
'Impreza'	46 419	35 844	49 114	55 136	57 410	60 282
'Forester'	–	–	3	37 730	80 455	81 441
Others	18 176	8 899	4 764	1 497	1 083	1
Total	160 235	154 247	183 278	229 779	274 080	274 191

Impreza Sti Version Sales Record (Domestic Market)

	1994	1995	1996	1997	1998	1999
'Impreza'	663	3 229	4 682	5 752	7 133	5 804

three goals that were established at the beginning of the campaign in the late 1980s. As a proportion of total domestic sales, sales of mini-commercial vehicles have fallen from 70 per cent in 1993, to less than 30 per cent of total unit sales in 1999 (see Table 5.3). The success of these strategies over a turbulent decade was no doubt one reason why, in December 1999, Fuji Heavy Industries announced that the American automobile giant General Motors had taken a 20 per cent stake in the company.

Conclusions

The isolation of one aspect of what has been a consistently focused and strategic plan for change is fraught with difficulty, if one attempts to ascribe the effectiveness of the plan to a single cause. However, in the case of Subaru, there is considerable evidence that the remarkable success of the motorsport campaign conceived by Fuji Heavy Industries and partnered and influenced by Prodrive, is a crucial aspect of the company's success. Indeed, the company's primary objectives were certainly achieved. The company succeeded in:

- Differentiating its products though a distinctive set of values and characteristics
- Giving the brand a consistent and unique image
- Successfully marketing the changing product mix from high volume, low value added vehicles to a set of distinctive, niche market, higher value products
- Maximizing the utilization of AWD, ECVT, TCT, Boxer engines and other technologies

The customer perceptions of the product, certainly in Europe and North America, have also been changed and several studies have shown that customers, actual and potential, have revised their image of the brand. These aspects of their strategy have certainly provided a strong brand image that is related to the powerful characteristics of SPEED, STYLE and SPACE. The company's product mix has also been successfully sold across different markets including conspicuous sales success in North America, where the rally campaign has not been dominant. The possible reason for that is the improved image of the vehicles with the high technology, speed and reliability of the rally cars strengthening the brand.

The company has also had considerable success in revising its model line-up and expanding the proportion of its output that is exported. In 1988 the company exported 45 per cent of its total production, including local production. The company's production at that time consisted exclusively of mini vehicles. By 1999 the export (exports plus local production) proportion had increased to 47 per cent but, more importantly, only 2973 units out of the total exports of 280 000 were mini vehicles. Furthermore, the proportion of mini vehicles produced by the company fell from 41 per cent in 1989 to less than 30 per cent by 1999.

In financial terms the company has had a remarkable record throughout the decade, being described by the Economist Intelligence Unit as 'Despite being the smallest vehicle manufacturer in Japan and one of the smallest in the world, FHI's financial position is stronger than some of its Japanese compatriots'. (EIU Automotive, Japanese Motor Business, 3rd Quarter 1998). The report emphasizes that the company's financial position has improved considerably in the last half of the decade, moving from a loss-making situation in the first three years of the decade to a strong profits and turnover improvement from 1994 to 1998. Gross profits and automobile sales have increased every year from 1995 to 1999.

The most successful element of the company's strategy over the past decade has probably been the manner in which the consistent success of the rally cars has been utilized to improve consumer perceptions and the technology of the model range. This is evident in the company's considerable sales success in the United States of America with the Forester model, a market where rallying is not important. However, in all markets the consistent evolution of the model range has been subtly influenced by the rally cars, both in terms of external design and technology. This has enabled the company to focus on ensuring that its model range, including the mini vehicles, are in niche market segments and perceived as high value-added products. This has been the basis of the company's market, sales and financial success.

6

Portrait of the Partner

The Prodrive Group

The Prodrive story began in the in the late 1970s, when the David Richards established D R Racing, to manage the Middle East Rally Championship under a consultancy agreement with the Rothmans tobacco company. Essentially the company acted as Championship and events organizer, ensuring that the brand derived maximum exposure from the events. By the beginning of the 1980 season, David Richards' partner Ian Parry was brought into the company to concentrate on the commercial development of the motorsport consultancy business. This was essentially developing the company's reputation and ability in assisting partners to obtain the maximum benefit from their motorsport expenditure. An objective that has never altered, but which has certainly become more focused.

By the end of the 1980 season, David Richards had persuaded Rothmans to support a Ford Rally Championship team with Ari Vatannen as driver and Richards as co-driver. The pair went on to win the World Championship in the following year, making Richards probably the most successful competitor amongst modern motorsport managers. The company continued to be active in almost any field or activity connected with racing and rallying, that would enable it to earn revenue and profit: in that order. Whilst Ian Parry continued to focus on the consultancy business, Richards retired as a competitive co-driver in 1982, but established the Rothmans Porsche Rally team in 1984 and also became involved in the March Company's Formula 1 activities. It was a period of very varied activities for the two partners and their fledgling company.

As David Richards has said:

'Those formative years were a period when we tried our hand at a wide range of business initiatives. In hindsight, we were like so many new ventures, and often failed to focus on our core skills'.

However, in spite of their rather diverse and somewhat scattergun approach to the business, a not uncommon feature in a fledgling company, they stuck to a number of fundamental principles which hold to this day. Indeed, this fledgling company was a very strong foundation, upon which they were to build one of the most successful companies in the United

Kingdom motorsport cluster, all out of a consultancy agreement with the Rothmans Tobacco company. The Rothmans Porsche 911 SCRS team provided the company with a European platform as well as continuing the company's activities in the Middle East.

The company was, and is, an unusual proposition, in that the dual strands of engineering and marketing have always been the dual principles and 'core competences' that have driven it, the company constantly focusing on the powerful combination that the two skills can provide for sponsors. Under the leadership of David Richards the company certainly lost little time in rejecting that somewhat familiar fragmentation of effort, and swiftly developed a rarely found focus in the degree to which it recognizes its strengths and core skills, and applies them.

The two men have very diverse backgrounds, and are totally different characters. David Richards is (as are many of the characters leading companies and the teams in this industry) the epitome of the 'high achiever', with a remarkable record as a World Rally Championship co-driver to prove his cool office management skills in an extraordinarily hostile environment. Ian Parry is the apparently laid back character with a sense of humour for all occasions, who is also the quiet, unassuming and meticulous manager who is always over-seeing contractual details, ensuring the specific aspects of any deal are settled unambiguously, and following up on the necessary fine print. The secret, or one of them, for their joint success, is that they have a remarkable set of complementary skills.

In the early years of the company the concentration on the combination of marketing and engineering skills was a natural recipe, given the skills of the founding Directors. Both had had successful careers in marketing and, in David Richards's case, also in the engineering aspects of rallying and race preparation. In this context the mission statement is extremely clear: 'To provide a commercial and marketing platform for the successful promotion of clients'.

In these early years the company was based at Silverstone and the founding partners managed the company with a rare financial conservatism, maintaining cash flow and profitability. In addition they were determined to maintain an 'open' approach to their company's financial strategy, always being able to provide the necessary financial probity to sponsors and latterly, investors. The Rothmans' link gave them the necessary associations with the large corporations involved in motorsport, and by 1984 they had re-formed the company into Prodrive Limited, the same year that they were running their first motorsport team, the Rothmans Porsche 911 SCRS Rally Team. In the first year the team had won the Middle East Rally Championship and secured second place in the European Rally Championship. During that first year the team scored a total of seven rally wins, a clear indication of the success to come. In the following (1985) season, the Prodrive Rothmans Porsche Team retained the Middle East Rally Championship and won the Irish Rally Championship. The success of the Rothmans Porsche 911 SCRS team provided the company with a European platform as well as continuing the company's activities in the Middle East.

The company's success in motorsport competition helped produce a level of commercial success that allowed them to develop a new facility at Banbury. That facility, which remains the company's headquarters today, is both visible from the M40 motorway, and houses an increasing number of people with high-level intellectual capital; a true knowledge-based company.

In 1987, Prodrive's successes in rally competition convinced the Bavarian motor manufacturer BMW that the company was the preferred partner to lead the company's World Rally Championship and touring car campaigns. In the first season of the World Rally Championships, Prodrive claimed its first WRC win with the BMW Rally car, in the Rally of Corsica. In addition to the successful launch of the BMW WRC Team, the company gained a total of five rally wins with BMW and Porsche, including second places in the Middle East and French Rally Championships. On the track the Prodrive BMW M3 contested five rounds of the British Touring Car Championship.

In 1988, the first full year of competition, the Prodrive-prepared BMW M3 won the British Touring Car Championship and the marque's rally successes continued. The Prodrive BMW rally cars won the Belgian Rally Championship and second places in the French and Italian Championships, and in the process gained a total of thirteen rally wins for the company, a record. In the following year the BMW Prodrive M3 won their second British Touring Car Championship and the rally cars won the Belgian and French Rally Championships adding a further seven rally wins to the company's record. By 1989 Prodrive were a conspicuous and very successful company in the field of developing rally and track cars to enable manufacturers to capitalize on the marketing benefits of motorsport.

It was at this point that the company's somewhat eclectic range of interests tended to reduce their financial success. As part of the company's determination to capitalize upon its rally and racing victories they had invested in a BMW dealership. However, the widespread recession that set in across Europe, but was particularly deep in Britain, damaged the sales of even BMWs as the customers themselves had to reduce expenditure and, more importantly, company car sales fell. It was certainly a clear signal to both Richards and Parry of the vital importance of focusing on the things that you do well and are consistently, sufficiently good at to ensure long term profitability.

The company had proved its effectiveness in the rarefied atmosphere of World Rallying and the British Touring Car Championship. In addition the company's commercial stability and management expertise were continually developed, as the company's founders determined to build a successful commercial and engineering team to ensure continued success. A decade later, many of the people who were employed at the start of the company remained part of a clearly outstanding organization. In the same way that Mr Yamada emphasized the vital importance of the 'soft management issues', the current Chairman of the company and founder Director David Richards has stated: 'I firmly believe that recognizing the talents of an individual within a team environment is one of the greatest management challenges we continue to face.'

The company has always developed talent, both in terms of the essential core of engineering expertise and in the driving seats of its race and rally cars. The engineering Director David Lapworth is one of many people in the company who have benefited from the fact that the company does, at least in respect of the Chairman's statement above, practice what it preaches.

By the end of the decade the company had achieved an enviable record of success in rallying, with Porsche and then BMW, and on the track they had won a further two British Touring Car Championships. A significant element in the next phase of the company's development was the initiatives that were successfully completed during the early years of the next decade, that began the long term partnership with Fuji Heavy Industries and their Subaru brand. That long-term partnership, detailed in the pages of this case study, proved to be an important factor in the way in which the company's strategy evolved. In particular the relationship has enabled Prodrive to take the essential and central skills of engineering excellence and marketing abilities, into other areas of the automotive sector.

By the end of the 1990s, after BMW, who really did not want to go any further in rallying, had sub-divided their British racing loyalties into the nineties (cutting Prodrive revenues), Prodrive and Subaru decided to form a partnership to pursue the World Rally Championship. The Japanese manufacturer had not been effective previously, but with first the Legacy and then the Imprezza range, Prodrive proved from 1989–99 that the Subarus they created and entered could enjoy regular World Championship success. The company consistently exceeded the Japanese company's expectations, winning three World titles for Subaru in the nineties, providing Britain *en route* with a World Champion rally driver for the first time.

The current Prodrive Subaru contract was set to run until 2001 and has weathered the loss of principal tobacco sponsor BAT's 555 brand, after they had decided to go Grand Prix racing with BAR-Reynard, at the close of 1998.

The Subaru business relationship has since been expanded, both into the development of yet more rapid road cars and an 'Arrive and Drive' rally team which can charge more than £1m a year to suitably qualified competitors. This particular offer is very attractive to the well funded privateer, those wishing to concentrate on driving rather than the considerable organization required to prepare, ship and service a modern rally car on far-flung events. Prodrive is reported to have supported over 100 competition vehicles and sent crews to more than 70 international events in 1998, when they held both Subaru 555 World Rallying and Honda Motor Europe Touring Car contracts.

The company was absent from touring car racing, before two years of wins in 1997–98, but gained no titles in the British and some selected European National Touring Car Championships with Honda. In 1999 Prodrive acquired the biggest spenders in the BTCC series, Ford Motor Company, and their racing Cosworth-engined Mondeos. To date the Fords have proved more competitive.

Another interesting set of circumstances occurred in the latter years of the decade was the involvement of David Richards in the Benetton Formula 1 team. This proved that Prodrive can operate without co-founder Richards and shows how well the company has avoided the usual British one-man entrepreneurial organization structure. Prodrive certainly fulfilled their promise of profitably increased turnover through much more work for its diverse divisions in engineering.

The company continued to build on the success it had achieved throughout the 1990s, through the development of new skills and ideas, within the context of adding value to the customer's motorsport campaigns.

Prodrive Now

Since its foundation in 1984, Prodrive has pursued its mission statement with remarkable success. In particular the company has consistently developed its ability to create and develop innovative solutions for clients, both in motorsport and the mainstream automotive industry. To achieve these objectives consistently, the company believes it has to employ extraordinary people and invest in the finest facilities and latest technology. The long partnership with Subaru is a clear example of their success in this endeavour.

By the end of the 1990s the company's market, technological and financial positions were at a point where a new set of innovations were probably required. The turnover had grown from slightly more than a million pounds at the formation to approximately £28m by the mid-1990s. Perhaps more importantly, the company's turnover had altered from being dominated by motorsport activities, to a position where a significant proportion of turnover came from its consultancy activities in automotive technology. By the end of the decade the turnover had grown to over £52 m of which over 30 per cent came from consultancy activities. More significantly, the company's profit record was rather more impressive as profits had grown from £1.6 m in 1996 on a turnover of approximately £29 m (approximately 5.5 per cent), to over £5.1 m on a 2000 turnover of £59 m, slightly over 8 per cent.

The end of the decade was a significant period in the company's development, in fact it was a period where its natural growth led to a re-evaluation of the strategy. For various reasons, the company had to undertake this detailed review, not least because it was becoming very clear to David Richards and his co-directors, that if the company was to continue to grow, an injection of capital was required. In 1999 the company's turnover had grown by 22 per cent over the previous year, and the organization was well prepared for the next phase of expansion through a detailed strategic plan, that was formulated in the last two years of the decade. The key elements of the plan were, firstly the partnership with American venture capital group Apax who purchased 49 per cent of the company's equity, and the separation of the two parts of the business.

The company's organisation is now spilt into two business streams, Motorsport and Automotive Technology. The Motorsport business based at the company's headquarters in Banbury runs factory teams in racing and rallying, as well as enabling customers and individuals to compete in rallies across the world. During its illustrious history, it has won four World Rally Championship Titles, and five British Touring Car Championships, the most recent being in 2000. In recent years, Prodrive has used technology proven in competition and its project engineering experience to offer the mainstream automotive market a design consultancy service. This business, called Automotive Technology, is now located at a 240-acre site near Warwick acquired in March 2000. This state of the art facility offers vehicle manufacturers and their suppliers everything from new technology development to niche performance vehicle design and build, as well as having the advantage of an extensive test track.

The company's continued and consistent success in motorsport has been replicated by considerable commercial success. This has been achieved by ongoing investment of resources in the latest technology and people to support the development of ideas and concepts generated by its designers and engineers. There is no doubt that the company's new investors recognized both the past success and the future potential of company and Apax Partners and Co., the private equity investment business, will certainly look for further growth in turnover and profitability in the future. Four company Directors: David Richards, chairman, Ian Parry, commercial director, Hugh Chambers, marketing director and David Lapworth, motorsport director, own the remaining 51 per cent which should ensure that the company will sustain its determined and continued focus on its mission statement.

PART III

THE STORY OF TWO CAMPAIGNS:
THE JAGUAR STORY

*The Building, Rebuilding and Rescue of the
World Class Jaguar Brand*

7

Jaguar, Jaguar Sport and Le Mans

Introduction

The association of Jaguar with motorsport success goes back to the company's roots in Blackpool, in the era of the Swallow Sidecar, and, continues today in the considerably more expensive and sophisticated World of Formula 1 Grand Prix. Not only dramatically more expensive but with the 'small' advantage that at least some of the expense is offset by contributions from sources other than the company, in this case Ford Motor Company of Dearborn, USA.

In 1924 Jaguar founder William (later Sir William) Lyons made a deal for a leading competitor to run a racing version of the Swallow 'chair' in the world-famous Isle of Man TT races. William Lyons found that, between practice sessions, the Swallow had been replaced by a rival brand from Birmingham. Typical of this type of man (there are currently no females involved in the management of motorsport with the departure of Nicola Foulston, but there undoubtedly will be), his short fuse was ignited and vocal repercussions followed. The Swallow was reinstated and ran to third place.

Sir William Lyons was a keen motor club member and competitor who always appreciated the value of international motorsport as a marketing tool, and used motorsport to help market his products, right from the start. However, as the best of Jaguar historians, the late Andrew Whyte, carefully noted, the experience had not been wasted on the man: 'Lyons had learned a lesson which was to make him forever circumspect in his approach to competitions as a manufacturer.'

Neither with its successful SS branded cars or the subsequent Jaguar marque, would their founder ever be prone to spending more than was necessary to achieve a headline result. The company publicly established the SS Jaguar brand in September 1935 for a motor show debut, and it was just two months before a major international competition was tackled by the marque. In the January 1936 Monte Carlo Rallye, an SS100 Jaguar sports car achieved second in class, but for Concours coachwork, not outright speed.

However, in 1936 the SS Jaguar achieved the kind of outright international rally honours they had been denied with earlier SS models. An SS Jaguar finished penalty free in the Swiss-based Alpine Trial (then a magnet for BMW and Bugatti) and took its first outright win overseas in Portugal the following year.

By 1938, Jaguar's sporting performance was beginning to be appreciated in what has always been its prime export market: America. In the United States a wealthy customer had taken his £395 SS 100 Jaguar for a season of US club events, mostly rallies and hill climbs in New England. This lone example of the Jaguar marque showed up extremely well in the company of then dominant MG sports cars and radically re-engineered Fords, wetting the appetite of a small but influential band of potential buyers, who were not able to receive left-hand drive (LHD) vehicles until 1947.

Back in the thirties, factory commitment from Coventry was usually limited to loaning cars to promising teams. These Jaguars were very simply prepared, but proved capable of 118 mph when timed at Brooklands. However William Lyons was enough of a competitor and salesman to also contact private owners when they looked as though there was no chance of winning against rival marques. The tough industrialist would not offer sympathy and encouragement, but asked those with no chance of enhancing the marque's reputation to withdraw and preserve the blooming Jaguar reputation!

The main story of the development of the Jaguar brand through motorsport, the marketing story, starts post-war, as does the vast increase in factory sales to the public (see Table 7.1). In the tables the deliveries were all cars consigned from the factory, as opposed to any used for internal purposes. Therefore, these are not total production figures as we have used totals for deliveries here in most pre-1980 contexts, and acknowledge a debt to former factory employee Andrew Whyte. Whyte diligently collected such statistics for his respected and widely published books on Jaguar topics, whilst the current Ford-owned Jaguar Cars Ltd supplied sales data from 1971 onward, completed for most major markets. In each tabular case the source is stated.

Pre-war, the highest Jaguar deliveries had been a total of 5378 delivered in 1939, but it took the introduction of LHD and the consequent benefits of an American and Continental European market opportunity to set the 'Export or Die' figures racing. By 1950 the wider range of cars available to potential customers (starting at a purchase price of £1247) had outstripped the pre-war sales record with over 6600 sold. A combination of developing export markets and the first of five Le Mans victories during the 1950s prodded new car deliveries over the 10 000 per annum barrier in 1953. Continued success on circuit and showroom saw output double again in 1959, enough to reach almost 21 000 units in 1960.

Table 7.1   Jaguar Cars:

Annual Deliveries 1950–70			Annual Domestic and Export Sales 1974–2000		
	Export	Total		Export	Total
1950	3926	6647	1974	12157	26632
1951	4273	5805	1975	12211	24469
1952	7978	8979	1976	14641	25042
1953	7643	10114	1977	12566	21953
1954	5335	10131	1978	12168	24980
1955	5438	9900	1979	9125	17160
1956	6847	12152	1980	9091	15011
1957	6614	12952	1981	9874	15562
1958	9177	17552	1982	15179	21619
1959	10476	20876	1983	22106	29175
1960	9677	19341	1984	25873	33147
1961	10174	24018	1985	29696	37745
1962	11135	22030	1986	33392	40971
1963	12235	24989	1987	35541	46643
1964	10206	24348	1988	34990	49495
1965	9560	24601	1989	33157	47400
1966	12098	25936	1990	32090	42754
1967	12026	22650	1991	19852	25661
1968	9800	24315	1992	16835	22478
1969	14970	27209	1993	21123	27338
1970	18422	30423	1994	23335	30020
Source: Andrew Whyte			1995	30929	39727
			1996	30579	39001
			1997	34251	43775
			1998	38550	50220
			1999	59829	75312
			2000	75028	90031

Source: Jaguar Cars 01/2001

What were the sporting activities that Jaguar attacked that obviously contributed to its powerful image? The pre-war Jaguar SS100 achieved the first post-war win in the 1948 Alpine Trial. However, it was not until Jaguar unlocked the record-breaking and racing potential of the voluptuous new XK120 in early 1949, scoring a 1–2 win at Silverstone, that their campaign took off. The classic XK120's sprint speed was complemented by durability, as Ian Appleyard proved in scoring the first of many international rally wins in the 1950s, with a 1950–53 Alpine Rally hat-trick. Other major rally wins included the astounding 1956 Monte Carlo victory for Ronnie Adams and two companions, in a Mk VII luxury limousine.

In addition to this consistent success, a young Stirling Moss was also winning races for Jaguar in production saloons, a tradition continued by later Jaguar production saloons into the sixties, and the company could have been satisfied. But Sir William Lyons knew the marketing value of motorsports, particularly when it was maintained and therefore provided continuous re-enforcement of the brand. The Jaguar racing manager F. R. W. 'Lofty' England was certainly able to deliver the continuous stream of major international successes, that were so necessary to provide a strong message to their most valuable of export markets: America.

The major breakthrough was with a development of the XK120, an open sports racing car known as XK120C within Jaguar, but immortalized as the C-type. This was Jaguar's first dedicated racing design, and it blended a production base from Jaguar's excellent XK twin-cam straight six engine (uprated from 160 to 220 bhp) to a tubular space frame. Former Bristol Aeroplane Company aerodynamicist Malcolm Sayer contributed the flowing aerodynamic lines of the bodywork that would became synonymous with more affordable showroom Jaguars.

A trio of 140 mph C-types contested that French pearl in any marketing man's modern strategy, the 24 Hours Le Mans race. In 1951 Jaguar fielded Peter Walker and Peter Whitehead, scoring a debut win in the 3.4 litre car. Even now, after a total of seven Le Mans victories, this first Le Mans win was possibly Jaguar's greatest moment in motorsports. F. R. W. 'Lofty' England commented, 'that was the event which set Jaguar on its path of success'. And Mr England was not a man who was easily impressed by events outside solid engineering progress.

The C-type also scored the second Jaguar victory at Le Mans in 1953 that was made more significant in the showroom, because Jaguar proved the motoring worth of Dunlop disc brakes. One of motor racing's most dramatic examples of 'improving the breed', and a very distinctive USP (Unique Selling Proposition) for any sports car manufacturer. However, it was Jaguar who was able to use the marketing advantage in all their models from 1957 onward. Jaguar's imprint upon the 24 Hours Le Mans race and its trophies, continued through the fifties. The company maintained annual deliveries above the 10 000 figure in every year except one (1955) after the 10 000 deliveries barrier was broken in 1953.

A further development of the C-type sports racing car, the single fin D-type and its aluminium monocoque, made it possible for the company to further develop its mainstay engine. The double overhead camshaft, straight six engine was continuously developed to give more power, throughout a hat trick of winning years between 1955 and 1957. Jaguar engineering enlarged the fabled XK six from 3.4 to 3.8 litres in 1957, a move that was soon echoed in production engines. This was yet another demonstration of the extensive links between race practice and the production models available in the showroom. For competition, this enlargement allowed 285 bhp, sufficient to push the car to a record 175.6 mph on the famous Mulsanne straight of the Le Mans circuit. These were heady achievements, but just as significant was the pioneering work carried out on the production engines. The enlarged 3.8 litre engine was initially made available for production cars, and then further developments led to a 4.2 litre, six cylinder engine that Jaguar continued to make until the mid-eighties.

Even after five Le Mans victories in the 1950s, a record of British success unequalled since the pre-war victories of the Bentley Boys, Jaguar would return to score two more wins. These came in 1988–90, this time in partnership with TWR (see Chapter 11).

Jaguar Cars Ltd officially withdrew from sports car racing in 1956. Their withdrawal was prompted by three factors:

1. The time constraints that racing and production developments imposed on an engineering team with limited resources. This was a vital factor, as the development team was responsible for all production as well as racing car development, as the public was demanding more advanced features at low cost.
2. The 1955 Le Mans tragedy, when 80 people including many spectators, were killed.
3. The withdrawal of Mercedes in 1956, who were then absent from the major motorsport arena until returning in the late eighties to meet established TWR Jaguars in long distance sports car racing. Commercially, none of Jaguar's present day competitors such as BMW, Audi, Porsche and Toyota-Lexus had then developed into the effective rivals they were to become in the late 1990s and the 21st century. So, in the late 1950s there was no commercial imperative to stay in motorsports, and Sir William remained wary of motorsports expenditure unless there was a specific cost *versus* value objective. As far as he was concerned motor racing was not to be pursued for its own sake.

Not everyone at Jaguar was totally happy to cease the annual quest to win Le Mans. Engineering director Bill Heynes commented to *Motor* magazine of motorsport's benefits to Jaguar:

'Far from being relieved when we pulled out of motor racing, I was more than sorry to abandon the competition cars. Not only had we achieved success on the track, but it had helped me build up a team of engineers in the design office, and on the shop floor, second to none in the industry. The enthusiasm which spread through the whole factory, I'm sure, made a great contribution to Jaguar success in production and in the World markets.'

Looking at the escalating Jaguar sales/production figures for the 1950s, it is impossible to disagree with that enthusiastic testament to the corporate spirit and commercial effectiveness of motor racing. The Jaguar generation of an in-house engineering elite and a boost to showroom traffic was also echoed in public statements by BMW AG nearly fifty years later in 2000, particularly when discussing their return to Grand Prix, and an increasing in-house manufacture of engine parts. Such practical sentiments as those quoted from Bill Heynes above, have also been echoed by senior executives from Ford on their return to Formula 1 through their acquisition of the Stewart team in 1997.

The private equipe Ecurie Ecosse, provided the marque with their last two D-type wins at Le Mans 1956–57, with ex-factory machines. These were to be the last racing successes for several decades, and Jaguar's spirited sales growth eased after their major motor racing effort ceased. In 1962, deliveries dropped slightly by 1988 units on the previous year, and did not exceed 25 000 Jaguar cars until 1966. The merger with the British Motor Corporation (BMC), later British Leyland, in July 1966 was subsequently admitted by Sir William Lyons to have been 'ruinous', although the merger preceded further production growth. Deliveries surpassed 30 000 in 1970, and a record was established in 1971 when 32 589 were delivered. This record stood for another disastrous decade of British Leyland ownership, as quality seemed to be consistently sacrificed for quantity. It was not until 1984 that sales surpassed that 1971 record.

The last racing successes of the D-type at Le Mans, were perhaps not as important for the public as the events involving another D-type which were taking place within Jaguar Engineering, at Browns Lane, Coventry. Firstly, Jaguar had created a production descendant of the D-type via an independent rear suspension prototype, known as the E2A. The consequent E-type redevelopment in soft top or closed coupe format was immediately perceived as one of the world's great sports or GT cars, a steel monocoque slashing construction costs compared to the cost of its aluminium hulled, racing parents.

The now legendary shape of the car and the 150-mph top speed ability of the E-type, coupled with an incredibly low price of £1843 in 1961, was the major UK sales factor. Worldwide, the obvious design links back to those Le Mans-winning motor racing cars of the 1950s, meant that Jaguar were happy to sell on the reputation gathered by their international motorsport honours, garnered in the previous decade. Perhaps for the same reasons, Mercedes were also in no hurry to return to the highest levels of motorsport at this point.

However, there were limited motorsport forays for the E-type, particularly to support the launch period. The 265 bhp/3.8 litre E-type straight six was re-deployed, as the E-type inevitably went racing against exotica costing rather more. The immediate impact, barely a month after debut at the Geneva Show in March 1961, was a televised Oulton Park race in near showroom specification. Graham Hill's brand new E-type beat Aston Martin and Ferrari opposition, and a second E type took third place, confirming that Jaguar still knew the advantages of motorsport for maximizing a new model's impact.

In 1962–3 Jackie Stewart, founder and consultant to the present Jaguar F1 racing team, kick-started a motor racing career that took him to three Formula 1, World Driver's Championships, in his brother's E-type. At the time he commented that the E-type was 'the most forgiving motor car I have driven ... You make it do anything you like.'

Although the factory did invest in the technical development of a low drag, aerodynamic body for the E -type by Malcolm Sayer, there was no official factory programme for the E-type to contest the Le Mans race. However, in 1963 the American enthusiast Briggs Cunningham entered a variant of the lighter E-type in the Le Sarthe event, bringing the car home in ninth place.

Jaguar did not return to motor racing officially until the 1980s with the XJ-S, and then it was a tentative start (the full story is covered by Chapters 8 and 9). Yet there were other competition programmes of commercial significance that Jaguar supported, with greater or less resources and enthusiasm in the sixties and seventies.

The compact 2.4, 3.4 and 3.8 litre saloons of 1956–68 were natural six cylinder competitors for touring car events. Jaguar also effectively loaned out their largest saloons to appear in British touring car races with winning results. Yet, the most effective commercial programme in Europe was that of the German importers, who raced the saloon cars in the European Touring Car Championship. Exports reached a record high in 1963, when Peter Nocker won the first European Touring Car Championship. The official Jaguar importer Peter Lindner shared the 3.8 with Nocker to win Germany's toughest annual touring car challenge, the Nurburgring 6 hours. This had obvious commercial attractions as the opposition was predominantly German, and Mercedes did allow many examples of their sportier saloons to appear in both races and rallies after their withdrawal from major motorsports in 1956.

Since the 2.4 litre Jaguar had also won at the dauntingly fast Spa-Francorchamps track in Belgium, the Continental reputation of Jaguar for making faster saloons at affordable prices was reinforced far beyond the reputation established by its Le Mans victories of the 1950s. As a result the export sales record of deliveries in 1963 was unmatched until 1969.

Some 127 380 of these versatile 2.4 to 3.8 litre saloons were manufactured over a period of twelve years. A much rarer production machine was the V12-engined, 2-door coupe version of Jaguar's award-winning XJ saloon series, the XJ-12C. Just 1862 were made between 1974–78. Nevertheless, it was on this somewhat heavy touring car that British Leyland pinned their hopes of restoring 1976–77 racing glory to the Jaguar name. The shameful period of Britain's declining motor car industry, presided over by Lord Stokes amongst many others, seemed to continue for many years and never stopped until the end of the decade. After Lord Stokes' departure from British Leyland, Alex Park succeeded him. In a surprising move for the time, the state-funded car division and its ex-Ford director Derek Whittaker, announced that Jaguar would make an official return to the race track under the latest corporate name, Leyland Cars.

Derek Whittaker knew that Ford had built a post-1962 sporting reputation for their 'Total Performance Approach', which was in defiance of the agreement made by the Detroit Motor Manufacturers, prompted by the obsession with safety, not to participate in motor races. Therefore, he defended Jaguar's return to motorsport in Leyland colours, purely on commercial grounds; 'We are going racing with Jaguar for exactly the same reasons as we do anything else in our business. To sell more cars and make more money.'

A good idea in theory, but the cars involved in the racing programme, which was sub-contracted, were dogged by reliability problems and, therefore, did more harm than good to the brand's image. However, in 1977 Leyland sources did report a boom in Benelux combined country sales figures following the Jaguar's racing appearances. A claim we could not substantiate from official sources in 2001.

Unfortunately, Leyland press relations were routinely bombastic and jingoistic, often making the terminal error of predicting that the opposition, principally non-factory BMWs, would be crushed. The over-optimistic marketing professionals were always prone to attach too much to the racing BL Jaguar's potential. A brace of 'Leyland Jaguars' were, eventually, prepared at unreported cost by the Ford, Triumph and Mini championship winning racing saloon car exponents, Broadspeed of Southam, in Warwickshire.

The owner and Chief Engineer Ralph Broad had never produced a losing racing saloon before, but in spite of frequently leading races, the 12-cylinder racing car developed from Jaguar origins struggled to complete the distance. The XJ12C/XJ5.3C project was simply too heavy and the permitted development time far too short, to permit the engineers to overcome these all too obvious disadvantages, in spite of the 500–570 horsepower advantages of its modestly enlarged 5.4 litre V12. The Broadspeed XJ 12C raced eight times, only once in 1976 and seven times in 1977, and never won a race. The best result, after terribly public engine and transmission failures, was an encouraging second place at the scene of previous Jaguar triumphs, the treacherous and demanding Nurburgring in Germany.

As in many highly competitive fields, the reality behind the failure was rather more dramatic and personal than the public realized. In addition to the public humiliation of losing, a circumstance that he was singularly unused to, Ralph Broad had to cope with the personal tragedy of suffering the loss of his daughter in a contemporary road accident. The public humiliations of the failure of the project, coupled with the tragedy of his personal loss, understandably overwhelmed Ralph Broad. The two-car team did not complete the 1977 season and the Broadspeed company never recovered, as it was taken over in the early 1980s, whilst Broad retired to Portugal. At press time the Broadspeed name was used in relation to a car import company that had previously specialized in retro Minis.

What lessons had been learned from this protracted debacle?

In fact British Leyland learned little about the mistakes they had made in either the management of the racing activity, or their commercial and Public Relations, often continuing in the same boastful tone. However, over in America during the same period, a lot was learned about the commercial values of racing the later XJ-S coupe, with the benefit of being run by Group 44 and Bob Tullius, a subject we will return to in a later chapter.

The overt arrogance of the British Leyland team is unthinkable today. When compared with the carefully spin-doctored statements of every driver and engineer involved in Jaguar's Formula 1 project, never mind management, it is quite apparent that lessons have been learned. However, in spite of repeated failures on the track, media debuts still seemed to seduce Jaguar employees and contracted team members into an embarrassingly high profile attitude. But the greater professionalism, and singular awareness of the implications of marketing, perhaps belongs more to the new Millennium world of major motorsport and the Formula 1 Grand Prix circus, as we shall see.

In 1984, the new Jaguar management team, led by Sir John Egan, had been successful in purchasing the once proud but then ailing company, from an even more ailing British Leyland. They were somewhat justifiably wary of supporting motor racing officially, in view of the very deep pockets such activity then required and because of the apparent ease with which failure and ignominy could be achieved. It would take a very tempting programme with the financial and prestige risks initially borne by outside contractors before Jaguar, with their new found freedom as a privately owned company, would dip their company paws back into the motor racing pond.

8

On Track to Post-Leyland Recovery

From Privatization to Rebuilding the Brand

In 1979, the official racing Jaguars were silenced, as British Leyland went through corporate agonies that threatened even Jaguar's existence, in the early eighties. Given what had happened to Britain's once globally dominant motorcycle industry, Jaguar employees faced the previously inconceivable notion that they could sink with the British Leyland ship, or simply be killed off by their ignorant owners. On the XJ-S front, sales were at an all-time low. In May 1980 John Egan became chairman of Jaguar under Leyland rule and he recalled:

'Do you know we were just a one product company then? For my first nine months we made no XJ-S coupes at all!' The Canadian importers ended that drought with an order for 100, Egan remembered, 'then we made another 100 and that seemed to saturate the market for a while.'

Over-production accounted for 1785 sales in that dire 1980 year, but there were real questions within Jaguar management as to whether it was worth continuing to manufacture the unloved XJ-S. The company had faith in Egan who, despite the inevitable job cuts, had caught the attention and loyalty of the working majority. There were 10500 employees in the company when Egan arrived, making just over 14100 Jaguars. By 1983 a leaner 8000 created 28044 Jaguars.

Interviewed by co-author Jeremy Walton in the winter of 1984–85, just a few months after the privatization of Jaguar Cars Ltd in August 1984, the new chairman of the company, John (later Sir John) Egan, acknowledged the considerable commercial value of the company's motor racing history. However, he harked back to the Sir Williams Lyons era in diagnosing one of the key problems that faced Jaguar at the prospect of re-entering endurance racing: 'There is absolutely no question of our preparing competition cars here, as in the past. We have kept 650 engineers busy here so far and we'd like 900.'

103

It was obvious that the working priority for in-house Jaguar engineers remained engineering safety-conscious luxury cars for the showroom.

Following the dreadful debacle of the Leyland-owned period of Jaguar's history, particularly the experience of the 1976–77 touring car racing season (see previous chapter), it was surprising that the newly privatized company would consider competition in that category. In spite of that, motor racing was the path chosen for the newly privatized company and the Jaguar was again to stalk the tracks of international motor racing. It was certainly considered that this was one way to breathe sales life back into the XJ-S V12 coupe design that had succeeded but not replaced (that is a task facing the 2004 F-type), the E-type. John Egan recalled, 'Back in 1980 we were selling 800–900 XJ-S types a year. In 1984 we sold more than 6500.' Much of the credit for that goes to the outside contractors who pressed Jaguar to let them race the XJ-S on either side of the Atlantic.

From a British perspective, the sporting success that led to subsequent satisfactory showroom and circuit results were a direct result of the alliance between Tom Walkinshaw Racing (TWR) and Jaguar. An alliance that continued to scale the heights of international sports car racing with two World Championships and commercially significant 24 hours victories at the legendary Le Mans circuit and Daytona, in the USA. However, it was actually in the USA that Jaguar's sporting reputation was reborn and there was a period when Jaguar had the luxury of teams on either side of the Atlantic fielding cars at minimal cost to the Coventry factory.

The American arm to Jaguar's revived racing ambitions traced its roots back to the Group 44 equipe, who traditionally worked with British sports cars in National races. Particularly successful with Triumph, a marque that did not survive British Leyland ownership, the Group 44 partner and driver Bob Tullius had mid-seventies racing experience of Jaguar's V12 powertrains in the E-type. A more ambitious plan was conceived for the XJ-S in 1977–8, and the Virginia premises that housed Group 44 (ironically now home to the BMW-backed Prototype Technology Group racing team), turned out immaculate XJR-coded sports racing cars of mid-engine V12 design between 1982–87. These were the cars that took Jaguar back to Le Mans in 1985 and won nine major American races, but Bob Tullius told us in January 2000:

'once TWR started developing their mid-engined V12 racers, I always knew our programme was unlikely to survive. There was no bad feeling, I remain on good terms with Jaguar today and I understand their reasons for selecting the TWR team as sole representatives after the 1987 season.'

Whilst Broadspeed and the larger XJC model were waddling into European trouble, Tullius was busy winning a class within America's prestigious 1977 Trans-Am series, winning half the events contested, with sponsorship branding costs split between Quaker State Oils, British Leyland and the Jaguar marque. In the following 1979 season, Tullius again won the title for himself, and also secured the manufacturer's trophy, making him a very welcome visitor at any Jaguar gathering for years to come. Mr Tullius was present as a retired racer, but active Floridian aviator, at the Year 2000 Launch of Jaguar's Formula 1 car.

The XJ-S did recover, to become Jaguar's longest-lived model, spanning 1975–96 and being made by the thousand in the nineties with US sales alone running at over 4300 in 1994. This remarkable resurgence was due to four primary product and marketing actions, one specifically allied to motor racing, and a third drawing on motor racing inspiration to distinguish the XJ-S from its competitors.

The first was the introduction in July 1981 of an HE (High Efficiency Model), which was an important milestone, and in 1983 John Egan explained its significance as:

'Not only did the world's first production application of Michael May's 'Fireball' combustion chamber principle transform the XJ's fuel economy, but also the detail changes we made to the exterior and interior specification improved its comfort and aesthetic appeal.'

In other words, the Jaguar regained the leather and wood luxury finish always associated with the marque's core values.

The second factor was that, from a rather low-key beginning in 1982 to the winning of the 1984 European Championship for TWR's Tom Walkinshaw, Jaguar gradually grew in commercial confidence alongside their racing success. Thus, the livery went from oil sponsor Motul and predominantly black, through to 1983's predominantly white with British Racing Green stripes. But by 1984 the more emotive British Racing Green and Jaguar stand-alone livery was being used, rather than 1983's split with original backer Motul, serving to underline Jaguar's increasing competition and commercial confidence. In 1984 Jaguar also won the 24 hours at Spa-Francorchamps in Belgium, shattering a monopoly of BMW victories that was a distinct help in emphasizing to the Europeans, that Jaguar were aware of quality and durability issues.

Thirdly, Egan listened carefully to American market needs as this was the production priority in 1981, exhorting and motivating the reduced payroll to ensure that models required for the US market in 1982, made their showroom debuts on time. A simple, but hard to achieve basic, that had constantly evaded the Leyland management in the past. A clever marketing strap line, 'thundering elegance', underlined the American link between XJ-S track performances of the seventies and the obvious civilization that had long been a Jaguar hallmark.

Finally, in 1991 Jaguar applied £50 m (then $84 m), to a comprehensive refit of the then teenage (16 year old) product, with the emphasis on enhanced quality and durability. It was a tangible benefit of the Ford take-over, and indicated just how determined the new owners were to turn their £1.8 billion acquisition around.

The Partnership of TWR and Jaguar

The story of Tom Walkinshaw Racing, otherwise known as TWR, is shorter in time than that of Jaguar, but equally dramatic. In 25 years since its inception in 1975, the TWR Group has grown from zero to an organization employing 1800 people with a Group turnover of over £300 m; all still privately owned by T. Walkinshaw. The founder Tom Walkinshaw is an accomplished saloon car racer who has turned to business with even greater success. The group's turnover is now mostly outside racing, which was reckoned to account for no more than 10 per cent of group turnover in Autumn 2000.

What did the company achieve before and after Jaguar?

What lessons did Jaguar learn from these track athletes?

Before we answer those, sometimes painful, questions, our panel TWR Highlights encapsulates how TWR was transformed from touring car specialists to one of the biggest automotive specialist Groups in Britain.

TWR Highlights: 1976–2000

1975: Tom Walkinshaw establishes TWR as an independent engineering company alongside his racing career. Walkinshaw had worked as a development driver for Ford on race and road cars, but was best known in touring cars after a spell earning his spurs in single seaters. Walkinshaw stopped driving at international level in 1985 to concentrate solely on business.

1976: TWR moves into a converted industrial unit at Kidlington, Oxfordshire. Walkinshaw then a winning contracted driver for BMW.

1977: TWR expands links with BMW and Mazda. Walkinshaw wins Britain's oldest race, the Silverstone TT, driving an Alpina BMW, ironically beating Broadspeed Jaguar before its home crowd.

1978: Backed by BMW, TWR develops 530iUS for Group 1 Production racing: Walkinshaw wins four races.

1979: BMW generously sponsor/commission for two years, the 'County Challenge' in the UK for the modified BMW 323i. Drivers include Nigel Mansell and Martin Brundle. Mazda UK/Belgium back expansion in touring cars with RX-7 rotary from TWR.

1980: TWR wins British Saloon Car Championship with Mazda RX-7. Second and final year of the BMW County Championship.

1981: TWR wins both the British and Belgian Saloon Car Championships, the Spa Francorchamps 24 hours (Mazda RX-7) and the Paris Dakar Rally (Range Rover). The Sportpart programme for Mazda takes TWR into the after-market business that the company previously explored with BMW 'Hallmark' special road cars.

1982: TWR wins the French Saloon Car Championship. The first year of association with Jaguar and the XJ-S backed by Motul sponsorship. The company wins four races, and is 2nd in the European Championship.

1983: The first season of official Jaguar backing for XJ-S, which comes 2nd in series with four wins.

1984: The TWR/Jaguar XJ-S V12 team wins the European title and the Spa 24 Hours Race: Tom Walkinshaw is Champion Driver. Jaguar commissions World Sports Car Racing programme.

1985: The debut of the Tony Southgate designed sports racing Jaguar XJR-6 Group C car at Mosport, Canada with V12 engine. The TWR Team wins six European races with Rover V8.

1986: The XJR-6 design runs with the Silk Cut livery and wins the Silverstone 1000 kms race.

1987: The most successful year for the TWR V12 Group C (XJR-8) car. Winning the World Championship with eight wins in ten qualifying rounds. The TWR company is established in USA.

1988: The team wins the World sports car title again, plus Le Mans and Daytona 24 hours with V12-powered XJR-9. TWR and Jaguar establish joint commercial ventures JaguarSport, producing £38 500 Jaguar XJR-S, modified road car for public sale. TWR also establish GM-Holden Special Vehicles in Australia. The company's expansion begins.

1989: No world class wins for V12 XJR-9/11, but the XJR-10 (V6, bi-turbo) wins three times in USA. The controversial XJ220 (V12 4x4 prototype, then rear drive V6 bi-turbo) is made for JaguarSport at a joint production facility from a site at Bloxham, Oxfordshire. TWR produces a 6-litre version of JaguarSport XJR-S, which replaces the 5.3 litre original.

1990: The XJR-15 (6-litre V12) is announced by TWR with obvious conflicting interests with the earlier Jaguar XJ220 supercar project. TWR-Jaguar V12s win at Le Mans and Daytona 24 hours, and the Holden V8 wins the Australian Bathurst.

1991: A third TWR Jaguar World Sport Car title, with the XJR-14 Cosworth Ford HBV8. A significant event as the first successful alliance of Ford, Cosworth and Jaguar brands. The XJR-16 won five major races in USA. The XJR-15 one-make series of 16 cars supports Spa, Monaco and Donington with a record $1 m final race prize fund. Tom Walkinshaw acquires 35 per cent of the shareholding of the Benetton Formula One Team and becomes engineering director.

1992: The new Aston Martin DB7 is designed by Ian Callum (now at Jaguar) and produced at Bloxham facility. The XJ220 is still in production, but sales slow.

1993: The last appearance of TWR Jaguar factory cars at January Daytona 24 hours: all three XJR-12s retire. Volvo commission TWR-850 racing programme in Britain. The JaguarSport alliance effectively ends with Aston Martin installed by Ford, owners of Aston Martin, at the Bloxham XJ220 plant. Production of XJ220 stopped after potential owners resort to legal action to avoid paying full £415 000 purchase price, nominating XJR-15 as a superior option.

1994: Tom Walkinshaw becomes engineering director of Ligier. Michael Schumacher wins the Formula One World Drivers Championship with Benetton. TWR wins the British Touring Car Championship (BTCC) with TWR Volvo features 850 Estates. Aston DB7 enters production and becomes the best seller ever for that marque.

1995: TWR and Volvo win BTCC races and sign commercial (Autonova) production agreement with TWR having a 51 per cent shareholding.

1996: Walkinshaw buys control of Arrows Grand Prix, leaving Ligier. TWR designed chassis wins Le Mans for Joest-Porsche Team. TWR-Holdens win the Australian Championship. A specialist HQ & Technical Centre is opened for TWR at Leafield in Oxfordshire. TWR retain Broadstone and Kidlington (engine development) in Oxfordshire. New TWR commercial offices opened in Japan & Brazil.

1997: Nissan R390 GT1 designed and developed by TWR enters Le Mans. In addition a homologated road variant is developed at a reported total programme cost of $38 m. TWR powered (Aurora V8) car wins Indy 500. The second win for Le Mans Joest-Porsche/TWR. Damon

Hill takes second place the Hungaroring GP for Arrows. Volvo S40 finishes 4th in BTCC, Detroit show debut for C70 Autonova co-produced with Volvo.

1998: TWR 3rd at Le Mans, with four Nissan R390s in top ten. Autonova commences full production of C70 and Convertible. The S40 wins BTCC driven by Rickard Rydell. Aurora V8 wins its second Indy 500 win. TWR Holden's win the Australian Championship and Bathurst 1000.

1999: Renault and TWR sign an agreement to create and make the Clio V6 through Autonova, Sweden. Manufacture of MultiRider buses and taxis with Volvo. Last year of British Championship with TWR-Volvo S40 placed third. New 'Virtual Reality' design facility is opened at the Leafield site and attracts industry customers.

2000: TWR now earns 90 per cent of all revenues outside motor racing, and has a total of 1800 employees. Arrows GP Team is the most visible TWR racing enterprise, along with GM-Holden Australia.

TWR and Jaguar: a Special Relationship

There would appear to be something about British life that breeds motor racing business entrepreneurs. Tom Walkinshaw of TWR is a product of a racing nation that has produced characters as diverse as Colin Chapman at Lotus, today's Formula 1 supremo Bernie Ecclestone, McLaren's Ron Dennis, Sir Frank Williams and latterly, Eddie Jordan. Those names are familiar through Formula 1, but Tom Walkinshaw's business success is of a wider nature. Walkinshaw's first big successes were in touring cars and being closest to Chapman, Tom could drive the cars he had engineered at an international level. Totally unlike Colin Chapman's Lotus days, Walkinshaw put business first and racing second. That is at least one reason why the TWR Group and its ten divisions now employs 'close to 1800' around the world and has turnover in excess of £300 m p.a., with 90 per cent of that cash generated by engineering activities outside motor racing business.

It was all very different in 1976. Walkinshaw replaced Gerry Birrell, as the Ford Motor Company development driver, with special responsibility for optional performance road cars, which proved excellent training for TWR's future business. Like Birrell, who was expected to follow Jim Clark and Jackie Stewart to Grand Prix, Tom was Scottish, but there the similarities ended. Tom remains one of the strongest characters in the harsh world of top line motorsports, physically and mentally tough. He does have a sense of humour and knows how to party, but winning at business takes pole position in his personal priorities.

From 1976–84 one of the authors accompanied TWR on some European Championships as they won across Europe with Mazdas, Rovers and Jaguar's monstrous V12 XJ-S cars. This experience gave us some insight into the sheer force of character that Walkinshaw often displays in adversity. We drove straight to the old Brno track in 1977 for Tom to practice the road layout in our Austrian rental Opel, which was not insured for Czechoslovakia anyway! As it got dark, Tom would only halt when a soldier stopped us. Tom argued his case for continuing track education brilliantly, until the soldier stuck a rifle inside our little yellow Kadett. Tom then told the military that it was all the fault of his passenger, and that he, Tom Walkinshaw, had been encouraged to carry on against his better judgement! They let us go, but the writer had to report to the police twice a day for the duration of our stay.

The Walkinshaw/Quester CSL won that Summer 1977 weekend, against the much more powerful Jaguars, and later that year Walkinshaw also won the British TT against Jaguar in front of the home crowd. Tom was pleased with a professional job, but his loyalty to Jaguar (his parents ran Mk2 Jaguars), his engineering expertise and Scottish shrewdness and insight, told him that there was a much better way to engineer winning Jaguars than the way Leyland Broadspeed had done in 1976–77.

Jaguar was *the* big opportunity, but before Tom could get the then under-financed Jaguar company interested in motorsports, there were racing and high performance Sportpart road associations with Mazda to complete. The Mazda experience had taught TWR how to prepare cars for and win 24 hour races and many other Belgian and British championship events. TWR made plenty of mistakes before that 1981 Spa win, the primary difference between them and other teams was that failure inspired a stronger determination to learn and win.

But Mazda also taught Walkinshaw and TWR how an association with a manufacturer could prosper well beyond the race track. The result was the Mazda and Sportpart TWR programme of body design and engine performance parts, which preceded the similar schemes that were behind JaguarSport. This was a predecessor to the logical progression to engineering and building complete road cars that the company subsequently undertook. TWR design engineers have created a large and varied range of concept and production vehicles for many of the larger manufacturers and the list traces an impressive pedigree. The company was responsible for the production Jaguar XJ220, the XJR-15, the Aston DB7 coupe and convertible, the Volvo C70 coupe and convertible and have inputs to many more on the confidential list.

The Kidlington premises, which was the home of the JaguarSport activities so crucial to the company's early development, and which currently houses TWR's engine-building activities under ex-Cosworth engineer Geoff Goddard, is no longer the base for the production of super cars. Illustrating the flexibility needed to survive in this industry, the facility originally housed two BMW 5-series saloons, South African Rad Dougall's Formula 2 car and, technician Eddie Hinckley, who typified TWR employees in happily withstanding two decades of high pressure employment. All this and a lot of empty, echoing, concrete, space.

Most of the work undertaken for the mainstream manufacturers is undertaken in strict secrecy. Although today, TWR acknowledge having worked with: Aston Martin, BMW, Ford, General Motors-Holden, Jaguar, Lamborghini, Mazda, Nissan, Renault, Rover, Saab and Volvo, it is likely that the list is even longer.

The economic and racing breakthrough for Tom Walkinshaw and TWR, was the association they created with the Jaguar sports racing car campaign. But it took a long time to get the Coventry company interested, after the disasters of the Leyland-Broadspeed campaigns. Finally, after a very carefully run promotional effort, Walkinshaw persuaded John (later to be Sir John) Egan to agree an assault on the European Championship on a 'no-win, no pay' basis. This was an extraordinary gamble that would have been a huge and very costly risk for the relatively young company. Of course the Motul backed 5.3 litre V12s had won against the BMW's 3.5 litre 635Csi, but only because TWR ignored the big Jaguar power potential in favour of racing durability; to win you first have to finish! Once TWR had established that BMW could be beaten in Europe, the support from the Jaguar factory swelled.

However, Tom Walkinshaw is not a man who gambles foolishly, if at all, and he is extremely capable of determining the engineering quality and racing potential of a car. It was at the Brussels Automotive Show in early 1985 that Tom Walkinshaw confided his view of Jaguar sporting prospects to a small group of journalists. Travelling in his company's latest TWR modified Jaguar road car which had pre-dated the JaguarSport deal by four years, he stated that: 'it really would be possible to beat Porsche in sports car racing.'

In the face of the facts of Porsche's dominance of sports car racing, this seemed an impossible dream, but none of us laughed. Tom Walkinshaw, like David Richards at the similarly successful Prodrive, had a long established habit of delivering on his hunches. In fact, at the time of his statement, the XJR-6 Group C sports racer was already being designed back in Britain. Those V12-engined racers, with a capacity of anything from 6.2 to 7.4 litres, took time to win consistently. When the XJR series did conquer Le Mans,

Daytona and annexed three World Sports car titles, all at a time when Jaguar was being readied for sale to Ford, TWR's status altered forever. Indeed, the status of the Jaguar share value also soared.

The phenomenal success of the campaign placed the Jaguar brand across the headlines of the world's press, gave the marketing people the opportunity to create lines such as 'the cat is back' and offered a remarkable bonus to Sir John Egan and his Board in the price of the shares. It also altered the profile and potential of TWR Limited.

The company was now in amongst the Big Players of the United Kingdom's emerging, high performance engineering cluster, with all the clients and the multinational deals you can see in the TWR Highlights *aide memoir* on pages 108–109.

Tom Walkinshaw no longer drives the press to lunch, or shares hire cars with scruffy journalists, but he always says an enthusiastic 'hello.' The man at the heart of this £300 m empire is still happy to talk about his latest project. As his interests have expanded so have the subjects of his passions, from serious rugby (he is a major force in the sport and a club owner in the UK), or the prospects for Arrows earning more Formula 1 points. He will also fearlessly express his opinion about what a fool you are! Then just when you get angry, the slow smile will spread across his craggy features. Mr Walkinshaw will then defuse you with a charm that is surprising in such a famously iron man. But then few have ever said 'no' to Tom Walkinshaw more than once!

9

From Track Stars to Major Commercial Players

The Marketing Benefits from the TWR/Jaguar Partnership

In utilising the expertise and experience of the TWR organization, Jaguar, as with Prodrive and Subaru, were determined not to simply seek immediate and short term media coverage. In exploring how the relationship between Jaguar and its racing partners was utilized to translate the partnership into a longer term marketing alliance with Tom Walkinshaw Racing (TWR) it is important to understand the similarities that emerge from the two experiences. As with the Subaru case, the profile of TWR's strategies and activities in the Jaguar campaign provides an interesting and illustrative set of lessons as to the necessary conditions of success in such brand development strategies. The evidence is in the impact that the 1982–85 racing programme for the XJ-S had on the marque's worldwide sales, which is followed by a chapter exploring the partnership's 1985–91 sports car programme.

The thrall within which the Jaguar company was held by British Leyland, was almost terminal, the handicap that the parent company imposed was destructive and practically destroyed the essence of the marque. Freed from the handicap of British Leyland ownership which ran from 1966–83, the company immediately became determined and ambitious to make the return to motorsports activities that had given the marque such lustre throughout the 1950s and before the 1966 takeover. This determination that motorsports were a key factor in achieving the brand's resurgence remained, despite the somewhat erratic Broadspeed/British Leyland assault on the European Championships seasons of 1976–77, which was too recent and too negative to forget.

The Jaguar management were confident enough about their independent future post-Leyland to look, not just at motor racing with outside sub-contractors, but also at marketing vehicles with a strong link to those motorsport ventures. First, they needed some success to obliterate the unfortunate European memories of the 1970s. Then the Coventry based marque could build upon that reacquired reputation to sell more Jaguars, a ploy that was more successful through the 1980s than in the recession hit sales seasons of 1991–93.

As is evident from the sales chart below (Table 9.1), it is obvious that the model Jaguar chose to serve them initially in motorsport was also their longest-lived Grand Touring model, the 1975–96 XJ-S. The effect was remarkable, for the combination of product upgrades under Egan and motor racing's high profile, saw XJ-S sales escalate substantially. Jaguar XJ-S volumes recovered from an all time low of 1199 in 1981 to ten times that figure in both 1988 and 1989, which were record years for the model and the marque.

Table 9.1   **XJ-S Sales and Product/Sport Actions, 1975–96**

Year	Product/Race Action	Annual Sales XJ-S	% of Total Jaguar
1975	Debut XJ-S V12 coupe @ £8 900	947	3.9
1976		2 625	10.5
1977	New GM 400 automatic transmission	2 611	11.9
1978		3 396	13.6
1979		2 352	13.7
1980	Digital electronic fuel injection John Egan appointed CEO, April 1980	1 760	11.7
1981	XJ-S HE @ £21 753, interior upgrades	1 199	7.7
1982	TWR European racing starts, 5 wins 3rd in European Championship	3 111	14.4
1983	XJ-S 3.6 coupe/Cabriolet @ £19 250 & £20 756 TWR 2nd, Euro Championship	4 808	16.5
1984	TWR/Tom Walkinshaw European Champions, win Spa 24 hours	6 028	18.0
1985	XJ-SC HE Cabriolet @ 26 995	7 510	19.9
1986		8 838	21.6
1987	ZF auto and sports suspension	9 537	20.4
1988	XJ-S full convertible £36,000 standard ABS. [TWR] JaguarSport XJR-S 5.3 £38,500 [inc.100 Le Mans editions]	10 284	20.8
1989	Driver's airbag USA. [TWR] JaguarSport XJR-S 6.0 @ £45 500	10 665	22.5
1990		9 255	21.6
1991	New XJ-S range, restyled. 4 litre AJ6+ Instrument pack	5 557	21.7

Year	Product/race action	Annual Sales XJ-S	% of Total Jaguar
1992	XJ-S 4.0 convertible @ £39 900	4 326	19.2
1993	Jaguar's 6.0 V12, enhanced security	4 995	18.3
1994	New AJ6 4.0 in coupe/convertible @£36 800 & £45 100	6 044	20.1
1995	XJ-S 4.0 Celebration, extra equipment models	5 802	15.0
1996	April 4, end production	3 763	N/A
Grand total		**115 413**	N/A

The Impact on the Market

Turning to individual markets, we can see that the combination of product enhancements and motor racing had a particularly visible influence on all Jaguar model sales in both the French and German markets. In Chapter 11, Table 11.3, page 145, detailed market analysis for five primary Jaguar markets is set out, but from Table 9.2 below, it is apparent that in France sales almost doubled between 1984–5, from a miserly 262 to a marginally more respectable 406. Jaguar sales then enjoyed a period of sustained growth to reach a profitable level of over 1026 units in 1987, rising to a record 2018 in 1989. The recession then started to bite and, in spite of the collapse of sterling in 1992, sales fell from then until 1996; their lowest for the decade at 594. After that year sales climbed back steadily to the record and very profitable level of over two thousand sales in France, but not until year 2000 (see Table 9.2).

Table 9.2 **Jaguar French Sales Performance, All Models, 1984–2000**

Years	Jaguars Sold	Years	Jaguars Sold
1984	262	1993	600
1985	406	1994	605
1986	512	1995	717
1987	1 026	1996	594
1988	1 515	1997	802
1989	2 018	1998	860
1990	1 906	1999	1 550
1991	1 200	2000	2 302
1992	750		

In Germany, Europe's biggest individual car market before reunification in 1990, the motorsport 'petrolheads' might well argue that even the fated Broadspeed programme had lifted public awareness of the Jaguar as an option to the prestige domestic marques of BMW, Mercedes and Porsche. This somewhat tenuous argument is substantiated by the sudden resurgence of sales in Germany in 1976, from the 1975 aggregate of 543 units to nearly four times that (actually a total of 1983), the following sales season!

Jaguar sales in Germany then remained somewhat erratic with a record 2011 in 1978, followed by a collapse of the market in 1979. It is obvious that other factors than marketing were at work in that year, probably local market forces, as sales of all Jaguar models fell by nearly 75 per cent to a bankrupting 508 vehicles in 1979. Sales remained erratic from that year until 1984, climbing back from 508 to 1527 in 1980, then 2510 in 1981, collapsing again to 845 in 1982. Then in the period from 1983–90 the company achieved steady growth, from 1220 annual Jaguar sales in 1983 to a sustained average annual sales level of more than 2000 models sold for the rest of that decade. This massive improvement culminated in sales of over 2400 in every year from 1988 to a record sales level of 2478 in 1990.

This huge improvement in sales in the very important German market, which is particularly demanding in terms of build quality, probably had little to do with any specific racing activity. Vastly more important was the enormous strides that the company had made in improving build quality and the speed of product and innovation, and ensuring that the potential customer was aware of the facts. These improvements undoubtedly under-pinned a sustained improvement in Jaguar sales from 1990, except in 1992 and 1994. Annual sales doubled from 1997's record of 3151 to an impressive 6497 in the last year of the millennium. See Table 9.3.

'A Doomed Product?'

Based on our company interviews with senior management of the period, it is safe to say that the XJ-S was a doomed product when John Egan took charge of Jaguar in 1980; as he aptly said, Jaguar had become a single product line company. The combination of racing success, improved build quality, product innovation and reliability of the road cars had, at least to some extent, revived Jaguar's reputation. This made it possible for the marque and its products not merely to recover its popularity, but to beat its all time sales record after nine production years. The very fact that this was achieved over such a relatively short period of consistently good management, is a stark reminder of the appalling damage that was imposed on the product and the brand by British Leyland.

Table 9.3   **Jaguar German Sales Performance, 1976–2000**

Years	Jaguars Sold	Years	Jaguars Sold
1976 (Oct 75 to Dec 76)	1983	1989	2410
1977	1589	1990	2479
1978	2011	1991	2146
1979	508	1992	1881
1980	1527	1993	2014
1981	2510	1994	1430
1982	845	1995	2525
1983	1220	1996	2512
1984	1938	1997	3151
1985	2350	1998	4059
1986	1852	1999	6090
1987	2184	2000	6467
1988	2407		

The XJ-S may not have been the most stunning car design, but the fact that a new car from one of the world's greatest brands, never sold more than 3400 units whilst still a fresh product in 1978, was a sad indictment of the commercial abilities of British Leyland. There seems little doubt that the XJ-S model (or the marque?) would not have survived, never mind blossomed to a record sales year fifteen years into its 21 year life, without a credible and well marketed racing career. In addition there was the 'halo' effect of acting as Jaguar's only sports car offering at the same time that Jaguar achieved so much in the world sports car racing arena.

Nobody at Jaguar would claim that it was all their idea to return to motorsports, in fact it was the vision of an outsider that brought them back into the fray. Jaguar's Sir John Egan finally bowed to the considerable persuasive powers of one Tom 'TWR' Walkinshaw in the early eighties. Their alliance, from 1982's backdoor support on the track to the full blooded sports racing campaigns of 1985–91, would lead not just to spectacular global motorsport success, but also to a range of jointly developed and financed road cars.

Not all these Jaguar-TWR sport-linked programmes were a success, indeed potential XJ-220 owners were involved in a legal action with the company over contractual obligations, as we discuss in the following chapter. Yet Jaguar rediscovered, sometimes earning a bloody corporate nose, in the intense negotiations with the somewhat tougher skills of Walkinshaw, just how to engineer and market cars with a vital dynamic appeal. The always perceptive and intelligent Egan recognized that, however tough in negotiating Walkinshaw was, he usually ultimately delivered on his promises. In the case of a marque and a brand trying to recreate itself this was an essential ingredient in the task that Egan and Jaguar's management had set themselves. They, and their customers, had had only too much experience of failed promises on everything from delivery timetables to quality, to realize the value of 'delivering what you said you were going to' or, in the Japanese way, 'exceeding the customers expectations'.

It was critical that 'Jags' appealed to a younger, more dynamic and wider clientele, than just their traditional customers, typified then by the ageing, male, golf club set. Even in 2001, the age of new prestige car owners was a subject of constant debate within Jaguar and their German rivals. Lowering the age that new car buyers entered the marque's customer base, was to be a prime motivation in Jaguar's future involvement in Formula 1 Grand Prix. This major programme started with the purchase of the Stewart Formula 1 Team, and the programme is the subject of extensive coverage in Chapter 11.

The first steps in reinventing the Jaguar brand in the early eighties were hesitant and rather basic. As Roger Putnam, then running Lotus's very successful marketing campaigns for their road cars, recalled:

'it is easy to forget just what a mess Jaguar was in. I was at Lotus and negotiating with John Egan to join Jaguar in the early eighties. I flew into Baginton (Coventry) airport and all I could see on the approach to landing was fields full of Jaguar XJ Saloons and XJ-S coupes. Up to their axles in mud and unsold. By then, I had experienced more than enough of Lotus and Colin Chapman's erratic genius. I wondered if I was not jumping from a small frying pan into a larger one!'

Roger Putnam continued;

'When I did join Jaguar as marketing manager, in 1982, the problems were appalling and many concerned the brand itself. Aside from the obvious lack of productivity, we employed 14000 and built 11000 Jaguars per annum, there was actually no custody of the brand at Jaguar! In fact John Egan had been very brave in hiring me, because the Ryder Report, which had led to the amalgamated Jaguar/Rover/Triumph Group, allowed for no sales and marketing activities based at Jaguar itself ... Jaguar had almost become British Leyland Large Car Plant number 2!'

Putnam's words were couched casually, but the deadly commercial message was clear in those crowded Baginton airport car parks. Potential customers had heard that Jaguars were not very reliable, tarred with the Leyland brush, and they were also not getting any message of Jaguar individuality; there was not one to give! The core values of the brand, essentially what Ford subsequently paid Jaguar some £1.6bn to acquire just seven years later, had been squandered. To the point at which the message from Michael Edwards to Sir John Egan on the latter's appointment to head up the Jaguar Board in 1981, was reportedly, brutally clear: 'fix it or close it'.

The overall marketing actions undertaken to rescue the brand and relevant to the motorsport and sales career of the XJ-S, were covered in our preceding chapter. However, Roger Putnam recalled some of the basic marketing measures taken as Tom Walkinshaw approached the still British Leyland owned Jaguar with that XJ-S motor racing plan. As Mr Putnam recalled:

> 'we began with the distribution and structure of our sales systems, but the most obvious action was, that we terminated agreements with 250 of the 300 UK dealers. They were only selling about 5000 cars a year between them and, for many, the Jaguar franchise was simply propping up a dealership that would otherwise have become insolvent. The dealerships that we did retain were extensively renovated and revised, with a corporate identity that once again branded Jaguar individually.
>
> Basically, we put the respect back into the Jaguar name, tried to turn it back into a complete car company as in Sir William's time. That was only possible because the product itself improved mildly through that HE variant, without costing Jaguar division money that simply was not available at this time.'

These actions had at least begun the long journey back to restoring the essential qualities of the brand, and the scene was set for other activity to restore brand credibility and respect, motor racing. Back in 1982, John Egan and his senior engineers warned Tom Walkinshaw to expect minimal public support from Jaguar for his XJ-S racing proposals. That was to be the case at least until the TWR team could prove that they had made the XJ-S V12 perform better than the tragic Leyland/Broadspeed Jaguar XJ-Cs of 1976–77. Undeterred by their cool reception at Jaguar, Tom Walkinshaw and his team set about restoring respectability to the Jaguar name on the race tracks of Europe, in a series traditionally dominated by Jaguar's bitter European rival, Germany's BMW marque.

TWR and their Oxfordshire based racing team achieved a total of twenty international victories with the XJ-S within the space of three years, between 1982–85, plus the 1984 European Touring Car Championship, with Tom Walkinshaw winning the Individual Driver's title into the bargain. There were also a number of victories in some of the world's toughest touring car races, including Belgium's Spa 24 hour race, and the 1985 Bathurst 1000 Kilometre race in Australia. In the latter event the XJ-S was actually competing beyond the formal agreement with Jaguar Cars, notably by separate agreement and finance from Jaguar in Australia and in the Asia Pacific Rim's Macao colony.

'You have to remember that Jaguar was still owned by British Leyland when Tom
Walkinshaw approached us'

said one senior Jaguar director, who preferred to remain anonymous. In a
further conversation he continued:

'the issue of cash spent on sport is always sensitive, and remains so under Ford
ownership. However, I can tell you that we achieved that European Touring Car
Championship, a win at the Spa 24 hours and many other worthwhile achievements
for annual expenditure in the £250 000 to £300 000 range. An all time bargain in our
estimation.'

It is important to point out that that those sums were not TWR's total
budget; they had other sources of sponsorship, including the oil company
Motul which was the primary source of revenue in 1982. Nevertheless, these
figures were a fraction of what a German company would have expected to
spend on such an ambitious sporting enterprise during this period. The TWR
track record of delivering results truly represents values that Sir William
Lyons would have approved.

Commercial Exploitation: The Development of JaguarSport

In spite of this record of success, executed so efficiently and effectively,
several important questions remain. How did the TWR-owned JaguarSport
company develop from what was a limited racing alliance that started in
1982? And what was the effect of that company's enhanced products on the
sales of the marque?

In the first instance the answer comes from Tom Walkinshaw's personal
experience at the Ford Motor Company during the 1970s. The redoubtable
Scotsman recalled that he had long been aware of the sales potential in
sportier versions of manufacturer's cars since his employment at Ford in
1973. At that time Ford made sportier versions of their mundane but best
selling Escort saloons, and extended these higher performance, often RS-
badged, variants to the Capri, their junior European version of the Mustang
'Pony Car' format.

When Ford's enthusiasm for higher performance variants waned in the
wake of the first fuel crisis, Tom Walkinshaw embarked on the creation of
what would become the present privately owned TWR Group. Firstly,
partnering with BMW and the specialists Alpina company, TWR's engineers
learned a great deal about the sophisticated application of transferring racing
attributes to showroom machines. The most visible result was the BMW 635

From Swallow sidecar to the 230 mph XJ220 represented 70 years of Jaguar's sporting business life. Throughout that 1922–92 span, Jaguar's belief in VFM (Value For Money) results was unshaken, but the year 2000 onward Formula 1 programme has yet to yield such value.

The two key figures in Jaguar's creation – and survival – as a billion dollar brand. Left is Sir John Egan (then simply John Egan, Jaguar Chairman) with Jaguar founder Sir William Lyons in the early 1980s. The products are the 1982 Jaguar XJ6 and a 1937 SS.

Think of Jaguar and the E-type typifies the marque's affordable high performance glamour. An image that rested not so much on the E-type's contemporary racing prowess, as upon the legendary Le Mans performances of the 1950s. For the public there was recognisable common ground between the racing Jaguars and the showroom E-type.

The XJ-S was planned as a direct replacement for the E-type, but muddled Leyland corporate thinking, and American legislation that appeared to forbid convertibles blunted its seventies sales debut. Motor racing the XJS with TWR restored the reputation of both Jaguar and XJS.

Under various managements since its 1966 sale, production has boomed and slumped erratically. Under Ford, consistent production increases have only been possible since the implementation of new product and quality programmes and later Jaguar products aimed at the more youthful market that is also targeted by Jaguar's move in Grand Prix.

The enhanced features of the JaguarSport XJR-S low production series – and that badge – were prompted directly by TWR-Jaguar sports success.

Jaguar scored multiple victories at Le Mans and built increased commercial success from that base. This is the first French victory in that annual classic, that of the C-type in 1953. In 1988 and 1990 TWR and Jaguar won Le Mans again to reinvigorate the brand during and after the Ford takeover of 1989.

Celebrating and commercializing their later Le Mans success with the XJR badged TWR-Jaguar racing cars, Jaguar and TWR cooperated in the showroom. Here the XJR-S formalises that heritage at Silverstone.

The Jaguar S-type of 1998 doubled Jaguar sales rates by 2000. Now the target is to double that figure again by 2005 with the addition of X-type with Formula 1 racing aimed at the younger market that is vital to the future prosperity of all prestige marques.

The Stewart-Ford Grand Prix HQ at Milton Keynes in 1999. The Jaguar Formula 1 team also operated from these premises in 2000 to 2001, but was scheduled to be rehoused along with Cosworth (motor engineering) and PI Group (advanced electronics) by 2003. All three companies belonged to Ford, part of a massive 5-year investment in Grand Prix racing.

Jaguar's Ford-branded Stewart predecessor as it appeared in its last season representing the Blue Oval. The Stewart-Ford won its first Grand Prix in that 1999 season, a feat well beyond the Jaguar successor during 2001.

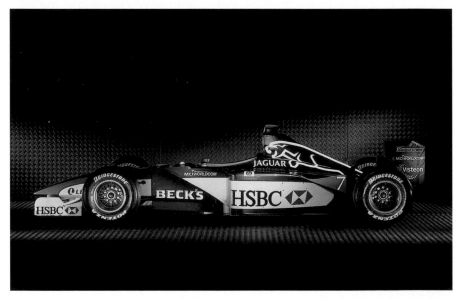

Rebranded with the high-profile 'leaper' Jaguar cat springing over its hindquarters, the 2000 Jaguar R1-Cosworth V10 brought public recognition over on-track success.

Rebranded with the high profile 'leaper' Jaguar cat springing over its hindquarters, the 2000 Jaguar R1-Cosworth V10 brought public recognition over on-track success.

Sportier showroom products tied in with a more aggressive appeal to customers in the 21st century. In this picture the XJ saloon typifies that commercial approach.

Sportier showroom products tied in with a more aggressive appeal to customers in the 21st century. Here, the supercharged XKR coupé typifies that commercial approach.

The 2001 X-type is aimed at the younger customers who have been targeted through the Ford-financed Grand Prix programme. Ford believed Jaguar virtually had to be in Grand Prix, competing against its natural showroom rivals from BMW and Mercedes. The X-type is expected to take Jaguar sales beyond 200 000 a year in 2005, but this is still small scale by the standards of the German prestige marques. Audi is the smallest of that German trio and it exceeded 600 000 sales in 2000, whilst Mercedes made over 1 million that year

CSi Hallmark, a low production series in the UK, that was especially useful to the company when it came to analysing the needs of Jaguar and their XJ-S, then one of BMW's biggest rivals in the luxury coupe sector.

Walkinshaw, increasingly through an ever-larger payroll of engineers, then revisited the mass market manufacturer Mazda, for another application of high performance engineering to a road car. The enterprise with Mazda proved particularly successful in the Belgian and British after-market role, and for TWR in a marketing technique link through Mazda. The greatest benefit for TWR and its growing revenues was the successful result of further negotiations, as Walkinshaw created a deal with General Motors' Australian division Holden, to create higher performance road versions of 9 m products, just as for Mazda. TWR then took that co-operation to its logical conclusion on the most suitable product line, the Commodore, to create the basis of a racing V8 saloon, to represent the Australian company in its local motor racing feuds with Ford Motor Company.

The GM-Holden co-operation continues to this day through a TWR Group-owned company Holden Special Vehicles; this has operated under both Special Vehicle Operations, and the current Holden Special Vehicles titles. This is not a small back street operation, but a substantial manufacturing and engineering organization, with the capacity to manufacture up to 5000 vehicles a year from its HQ in Clayton Victoria, a site where Nissans were once assembled for Australia.

TWR and JaguarSport Sales

TWR were proactive in trying to transfer some of the some of the reputation and performance earned by the racing XJ-S, into profitable showroom activity. At the Brussels Motor Show of 12 January 1984 they displayed a new road model 'with the full co-operation of the Jaguar factory', a £34 701 XJ-S. This car had a number of engineering features on both the engine and the suspension that offered improved performance, due entirely to the company's experience with the XJ-S on the race track; racing had improved the breed.

The most notable improvements were a 'blueprinted V12 engine of 5.3 litres', rebuilt by hand with more precise tolerances for maximum power, and incidentally much smoother power delivery. This improved power plant was mated to a 5-speed, manual gearbox that was not offered by Jaguar factory. Further specification changes embraced new brakes, enlarged in the light of racing experience, aerodynamically enhanced body panels, and revised suspension, that was considerably more responsive than that on the standard road car.

Always aware of the true essence of the Jaguar brand, TWR did not supply a version of the racing XJ-S, a deviant road car, but trimmed the complete package in wood and leather, replacing the steering wheel with a reminder of the TWR track lineage. It was a subtle upgrade of the XJ-S model's performance and character, which enhanced the brand's characteristics and delighted the customer. This was an important example to the Jaguar Company that Tom Walkinshaw and his team understood the demands of commercially successful co-operation, an essential ingredient for the manufacturer in such programmes.

The TWR JaguarSport XJ-S demonstrated, through its wide public acceptance, that there was a more luxurious, higher performance, market that a suitably modified XJ-S could satisfy. Significantly, TWR chose Belgium as the neutral ground for the models' launch, emphasizing that 'these cars incorporate many features proven over two successful seasons of racing in Europe', rather than taking an isolationist UK-only sales stance.

Throughout most of the 1980s, the TWR company continued to market their own TWR JaguarSport models, complete cars or individual component items, until the end of 1988. Unlike the production results of the formal alliance between JaguarSport and Jaguar, no complete record survives of how many were built. Primarily, this is because only the later, oversize 6 litre and manual gearbox option was built at the TWR facility at Kidlington, Oxfordshire. Unless models were provided directly from Kidlington, dealers dealt with the customers, who might opt for specific TWR modifications/ improvements, or more radical complete car conversions.

In 1988 the company achieved more racing success with Jaguar, which was celebrated by a more formal liaison with the manufacturer. Instead of a blessing from Jaguar for TWR to go its own sales way, Jaguar and TWR formed a joint venture, which was announced in May 1988, and the first product announcement, of a further uprated XJ-S, was made to press and public in August of that year. The allocation of shares in JaguarSport, was never made public, but TWR subsequently held 51 per cent of the further joint venture to build a factory and the XJ220 (see Chapter 10). Based on another building within TWR Group's original, Kidlington Industrial Estate home that is still used today, but housing the Arrows 3 seater AX3 Grand Prix demonstration team, JaguarSport's declared objectives were to: 'develop and manufacture sporting and high performance derivatives of the Jaguar range of saloon and sports cars.'

The company's statement on 22 August 1988 continued:

'The company is planning a start-up production level of 500 cars per year rising to 2500, dependent on customer demand. Initially sales will be in the UK market and 20 specialist JaguarSport dealers have already been appointed, all sourced from within the Jaguar Cars franchise network. Sales will expand into European and North America next year.'

Since total production of XJR JaguarSport variants failed to exceed 2400 units over six years, the highest production year is unlikely to have exceeded 500 units a year. TWR supplied only totals over several years of production for each model, and Tom Walkinshaw declined an invitation to comment on any aspect of the JaguarSport era.

TWR's American ambitions had to be curtailed in the light of the US emission and other Federal laws, but JaguarSport continued to supply both saloon and coupe versions of the XJ lines. As we have seen, these TWR assembled, XJR variants boosted overall XJ-S sales figures, more by association than showroom statistics, to the point where the XJ-S sold most strongly in its closing decade. Essentially the enterprise increased the model life of an increasingly important part of the Jaguar model range, thereby enhancing the brand and the company's profitability.

Sir John Egan, chairman of the JaguarSport-TWR venture, acknowledged the debt the company owed to the motor racing results that were achieved in the 1980s.

'We formed JaguarSport, to produce Jaguars that would appeal more directly to motorsport enthusiasts. I am confident the XJR-S will do just that. Winning Le Mans again gave us the idea of a celebration version (the first 100 JaguarSport XJR-S types were to this specification) and I am delighted with the end result.'

In August 1988, these £38500 packages reminded the public of that June 1988 Le Mans triumph, Jaguar's first since the 1950s, by utilizing the showroom name XJR-S. This is a play on the XJR prefixes of the sports racing cars backed by Jaguar, Group 44 and TWR.

A year later in August 1989, JaguarSport announced mechanical changes to complement the largely cosmetic enhancements of the 1988 XJR. Again, the name XJR-S was used, but this time the public were offered a 6-litre version of the charismatic V12 engine that enhanced the power output by some 17 per cent, and took the stated horsepower comfortably above 300 bhp. Racing links were quoted as:

'the car has been equipped with Zytek sequential injection digital ignition engine management system which has been directly developed from Jaguar's Group C racing programme. Zytek systems are also used on several Formula 1 cars.'

However, the XJR-S remained less radical than TWR's original public offerings, lacking the manual transmission in favour of the automatic version that Jaguar customers almost universally chose in V12 XJ-S sales.

The more powerful XJR-S was released in September 1989, and incorporated an even longer list of luxury features to culminate in a substantial price hike to £45 500, in the UK. It is apparent that both Tom Walkinshaw and Jaguar were now more bullish about Jaguar's status in the higher performance (and profit margin), leagues of car manufacturing. The extent of their optimism can be gauged from Tom Walkinshaw's contemporary assertion:

'the new XJR-S puts us very much on the pace with competition such as Porsche and Ferrari. It illustrates very well our policy of producing cars which are amongst the fastest and most exciting in the world.'

In retrospect the TWR-Jaguar alliance was at its most successful in the period from 1988 to 1994, when the span of the XJR variations on an XJ-S V12 coupe or 6-cylinder XJ saloon theme were sold. The most popular single model was just the 1990–94 XJR saloon, with 700 made and sold in a four year production life. Originally introduced in 1988, as an £8000 extra cost 3.6 litre, and then uprated to the factory's 4 litre specification, the TWR improvements to the factory XJR saloon were primarily focused on suspension and appearance changes. The XJR saloon version did not share the fundamental engine power increases that TWR engineered for their XJR-S coupes.

Production and sales of all JaguarSport products ceased in 1994. In the case of the mass production based XJRs, they had served their purpose to heighten awareness of Jaguar's success in sport to a new audience. More pertinently, the mass-produced Jaguars on which they were based, were coming to the end of their production lives, the XJ-S replaced in 1994 by XJ6-X300 and the XJ-S by the XK8 in 1996. This made it even more remarkable

Table 9.4 **JaguarSport production at TWR Kidlington, 1988–94 (XJ-S coupe and XJ saloon-based cars)**

Model, litres [bhp]	Years of Manufacture	Total Made
XJR-S 5.3 V12 coupe [300 bhp]	1988–90	350
XJR 3.6 saloon [221 bhp]	1988–90	400
XJR-S 6.0 V12 coupe [318 bhp]	1990–92	680
XJR 4.0 saloon [223 bhp]	1990–94	700
XJR-S 6.0 V12 coupe [333 bhp]	1990–94	250
Total, all types	**1988–94**	**2380**

that the XJR 4.0 litre saloon set the highest JaguarSport sales total, even whilst running into the last year of its production life.

Of a total 2715 JaguarSport products developed and manufactured in this period, no less than 2380 were XJR types (see Table 9.4), the remaining 335 of the exotic XJ220 and XJR 15 breed, described in the following chapter.

10

Recreating a £1.6bn Brand through Sponsored Motor Racing 1986–1990

On the Sponsored Road to Brand Pride and Ford Ownership

As Jaguar slowly gained confidence in TWR's ability to deliver the racing results, and in their ability to build and market a more appealing product, the Company became more ambitious again, both on the circuit and in the showroom. Having achieved their European objectives with the XJS, and shown that they could beat BMW in the Bavarian company's favourite touring car playground, wider and more ambitious plans were laid. The company's main objective was international sports car racing, and with it, a chance to expand their promotional presence in the largest market of all, for Jaguar and most other prestige brands, the United States of America and in excess of 40 m potential customers.

Sports car racing was, and is, a hugely expensive activity. Whilst the favourite sponsoring activity of many companies is currently ocean yacht racing, described as 'tearing up twenty pound notes whilst under the shower', sports car racing probably beats it hands down in terms of cost. Jaguar, in spite of their justifiable confidence in their new found partner, needed an injection of funds to pursue sports car racing as an effective marketing tool; that meant winning premier races such as Le Mans. Undaunted by the immense problem, another and crucial partner appeared for Jaguar, the independent entrepreneur Guy Edwards. He was able to tap in to an adequate pool of money for an efficiently sponsored effort, typically from Silk Cut Cigarettes in the World Sports car arena and Castrol in the USA and for Le Mans. The level and consistency of the sponsorship he delivered made it possible to continue the traditional and pragmatic Williams Lyons mantra for any racing activity, 'Maximum Value, Minimum Cost Per Achievement'. Perhaps the only realistic approach to expensive motorsports. But then almost all commercially attractive motorsports are very, very expensive.

It was not all good news, as the commercial partnership with TWR was effectively dissolved, after the XJ220 supercar ceased production following a commercially controversial five year (1989–94) production run of 285 examples, a subject with which we close this chapter. The production of this 'supercar' was perhaps one of the few ventures that TWR undertook in collaboration with Jaguar that failed to fulfil the anticipated targets.

The extensive sponsorship funds generated by Guy Edwards QGM (Queens Gallantry Medal), made the extended campaign to win World Sports Car Championship a financial possibility. But, more importantly, through providing much of the financial resources necessary to turn the plans for that venture into reality, indirectly made it possible to lift the price per share paid by Ford for Jaguar ownership in 1989. The XJR prefixed sports racers, also added vital victories to support the marketing cause of an ageing model line, whilst new models were being developed, following the Ford take-over.

What was Achieved?

The second stage of the alliance between TWR and Jaguar was the most important, certainly in the context of the continued development of the Jaguar brand, the first stage having achieved the brand's re-establishment. In the first phase of their partnership, Jaguar and TWR had forged the essential elements of trust and mutual respect, precisely as Prodrive and Fuji Heavy Industries had within the first two years of their relationship. In both cases each partner in the alliances had determined the boundaries of responsibilities, and sufficient success had been won for the commissioning partners, Fuji Heavy Industries and Jaguar respectively, to have confidence that their objectives would be met. However, in the case of Jaguar and TWR the commitment had evolved over the first phase of a rather looser alliance, whereas Fuji Heavy Industries had always been very clear about their aims and objectives, and had been very specific about the boundaries of responsibility. There are no doubt specific reasons for the differences beyond the culturally diverse factors within the companies, and we discuss these in the concluding chapter.

Therefore, in the second stage of the TWR and Jaguar motorsport alliance there was a strong emphasis on further developing the sporting pedigree and essence of the brand. The level of competition, the racing record and, necessarily, the sports racing cars that were created to further the XJ and XJR Jaguar sub-brands became more exotic and sophisticated. In addition the levels of risk and the costs were higher as was the level of commitment of both companies.

Building on the previous success with Jaguar V12 products as described in our earlier chapters, America's Bob Tullius obtained backing to return to Le Mans on an exploratory basis. That was in 1983, with the purpose-built racing

XJR-5 and a carburated, 6-litre Jaguar V12 engine. But the XJR-5 was an old fashioned and rather heavy design by contemporary European standards and the car's performance, whilst creditable, was probably not capable of competing against the might of Europe.

These sporting XJRs were superbly presented, in a predominantly white Jaguar livery and cost Jaguar North America money, rather than the cash starved United Kingdom parent company. Therefore, Jaguar Chairman John Egan was happy to see Tullius represent the company's sports car racing interests, including a return to Le Mans in 1984.

American Sales and Sport

The USA had traditionally been Jaguar's number one export market, but that did not mean it had escaped the British Leyland ownership effect. The lowest point, and there were many of them worldwide, was 1979–80, when less than 4000 Jaguars were shipped across the Atlantic. John Egan's insistence that the improved HE models of 1982 made the US showrooms on time, plus some excellent Trans Am Halo effect from Tullius's winning work on the race track with XJS in 1981, had a dramatic combined effect.

Sales leapt from just under 4700 in 1981, to more than 10300 in 1982. Jaguar's performance continued to improve dramatically on the track and in the showroom, as sales doubled again to more than 24400 by 1986. Recessions and ageing products, of sometimes dubious build quality in the earlier 1990s, had their effect on subsequent sales patterns. Sales of Jaguar models in the United States of America slumped to less than 10000 units annually in 1991, failing to top 20000 annually again until 1999. However, by 2000, the availability of the S-type, together with the improved Jaguar range, led to sales breaking that 1986 record, reaching almost 44000 in 2000. (See Table 11.2, Chapter 11.)

America Pioneers British Victories

The marque's success in the sports racing car arena in America was the foundation upon which the assault on the European contests would begin. Bob Tullius recalled in January 2000, at the debut of Jaguar's Formula 1 car:

'we knew that TWR in Britain would come with another development of the sports racing Jaguar to race at Le Mans. Nobody hid that from us. We just did the best job we could.'

The Group 44 sponsored Jaguars were fielded at Le Mans in 1985, having already shown winning form with the Lee Dykstra-designed XJR-5 in the USA, form confirmed with victory in the Daytona 3 hours with their XJR-7 at the close of 1986. The Tullius V12 delighted Jaguar enthusiasts by leading at Le Mans, on the marque's return to the track that had provided them with so much glory in the 1950s, but the car eventually finished a troubled 13th.

However, Tullius accepted the inevitable and, having raced at Daytona in 1988 without backing from Jaguar, the American operation was closed down. Ironically a BMW racing team acquired the Winchester Virginia premises shortly after the operation closed down.

The switch of Jaguar's factory sponsorship, from the XJS to worldwide, and later major American, sports car racing programmes, was made after an exploratory outing at the end of 1985, at Mosport, Canada. 'Exploratory', because TWR's detailed examination of the American Tullius/Group 44 sports racing machines, revealed that the new racers TWR would have to make very significant technical advances if the new cars were to be competitive in the European Sports racing car arena. The Tullius/Group 44 era car had been based upon a reinforced aluminium chassis which, whilst it had been successful, was not considered by the TWR engineers to offer sufficient development potential. The new design was to utilize the same technology as used in Formula 1 Grand Prix cars and aerospace technology, that is a carbon fibre monocoque chassis.

The new cars were running by the 1985 season and from then until 1991, TWR competed in assorted World and American sports racing programmes. These secured publicity for the Jaguar marque and, rather more importantly, three World Championships, plus two victories in the 24 Hour Le Mans in France, and two victories in the 24 Hour Race at Daytona in the United States of America. These four victories were the marketing person's dream for the future consolidation of everything that had been so hard-won in improving the brand.

The TWR team took up the XJR prefixes and created an official Jaguar sports racer, the 1985 XJR-6, designed by the highly talented Tony Southgate. Thus, a British-based Jaguar challenge with V12 power would appear at Le Mans from the 1986 season until 1990. The sheer power of 650 bhp from 6.2 litres, housed in a carbon fibre chassis beneath the purposeful looking British Racing Green body, was a formidable, and patriotic, attraction. However, there was little hope of Jaguar being able to afford to go racing in the highly emotive British Racing Green, bereft of major sponsorship. It was for this very simple reason, as it usually is, that the highly attractive Silk Cut and Castrol sponsorships came to be the livery of the famous cars (see section below 'The Modest Price of Success').

The change of designer and the technological developments achieved by the TWR engineers and designers were so successful that, by the 1986 season, the TWR XJR-6's winning run began, against the dominant turbocharged Porsches. Their first victory was, most appropriately, at Silverstone on home tarmac, the cars racing in their Silk Cut colours. The Walkinshaw sports racing V12s, the XJR-8, overpowered their opponents from 1987 onward, and in that 1987 season the Jaguars won both the drivers' and manufacturers' World Sports Car Championships, a feat they repeated in 1988 and 1991. In total the exceptionally competitive and reliable sports racing cars won six FIA World Sports Car titles between 1987 and 1991.

In spite of their early success in the 1987 Championships, the TWR V12 Jaguars had to wait until the following season, for the first of two Jaguar V12 wins in the most prestigious endurance races. The crowns of the Le Mans and Daytona 24 Hour races, were both won in 1988, and again two years later in 1990. Driving an aerodynamically elongated XJR-9 LM capable of 240 mph, Andy Wallace, Jan Lammers and Johnny Dumfries made history in 1988, taking the winning Jaguar to the marque's sixth Le Mans win, and the first for the company since 1957. The TWR Jaguar XJR-12 won at the Sarthe circuit again in 1990, with an update of the thundering 750 bhp, 7 litre, V12 formula in 1990. Thus was captured the marque's seventh Le Mans, and a tremendous 1–2 victory, to become the third most successful manufacturer, in terms of outright wins, ever to compete in the gruelling 24 Hour Le Mans.

The combination of TWR and Jaguar nearly won Le Mans again in 1991, finally taking the chequered flag in a magnificent 2-3-4 formation; perhaps coming second can be better than winning? Then, the massive Silk Cut V12s of 6 litres in the USA and 7.4 litres at Le Mans, conquered the Daytona 24 hours for a second time in 1990, despite some steamy interludes. Jaguar and TWR were forced to augment the legendary V12 power sources of such marketing potency, when linked to showroom products, to meet differing national regulations. That meant exploring leading edge turbocharging and Grand Prix technologies in both the 1989–91 USA and 1991 World sports car racing seasons.

In the American Camel IMSA GTP category, a 3.0 litre twin turbo V6 XJR-10 of 650 bhp, which was a distant cousin to the V6 utilized in the production XJ220, proved six times a winner in shorter events. Based at Valparaiso, TWR Inc ran a very effective five year campaign of racing missions in America from 1988–93, including a debut Daytona win, but both Jaguar wins were with the V12, suitably linked for print media advertising campaigns.

The V12 XJR-12 remained the backbone of Jaguar racing power, whenever durability was a priority over sprint agility. The 1989–90 missing link was the 3.5 litre XJR-11 twin turbo, sibling to the XJR-10. However, when turbocharging was banned in 1991, it had never fulfilled its full world sports car potential. Globally, Jaguar pummelled both Peugeot and Sauber Mercedes with the XJR-14, designed by current Ferrari Formula 1 genius Ross Brawn on Formula 1 principles, for the 1991–92 Group C and IMSA. The flyweight XJR-14 won three World Championship events by embarrassing margins, and both drivers' and manufacturers' titles in the Group C category.

The XJR-14 featured an adapted Cosworth racing power unit, in a preview of the present Grand Prix alliance between Jaguar and Cosworth. Albeit Northampton supplied a V8, whilst today's 800 horsepower Formula 1 Jaguar Cosworth, is of the universal V10 configuration. In the 2000 Formula 1 season, Jaguar's Cosworth 3 litre, Formula 1 engine, was also overseen by one of Jaguar's most experienced engine engineering executives, Trevor Crisp. We discuss that in more detail in the next chapter.

The Modest Price of Success

Jaguar's 1986–91 sports racing programmes, discussed later in this chapter, were also extremely affordable by corporate standards. The Jaguar Company only needed to contribute approximately one fifth of the annual £5 m plus budget. Outside sponsors contributed the balance, as a result of 'the incredible deals', that the entrepreneur Guy Edwards concluded, according to Jaguar Director, Roger Putnam. Speaking to us in temporary offices at the Coventry HQ, during a rare break from the early 2001 pre-launch preparations for the debut of the X-type at Geneva, Mr Putnam amplified on the astonishing value Jaguar Cars received. His remarks covered six seasons of TWR Jaguar sports car racing programmes in the five seasons that followed an exploratory Canadian outing in August 1985, from 1986 to 1991.

Unlike the North American racing programmes with Bob Tullius, which were financed by Jaguar North America, Mr Putnam was the self-described 'paymaster' to the TWR Jaguar operation for those sports racing years (1985 to 1991). That meant collecting all the sponsorship fees, the Jaguar company's contributions, and then paying TWR's fees. The Walkinshaw equipe were responsible for all racing matters, from car creation to racing the V12, V6 turbo, and Grand Prix (HB) Cosworth V8 powered XJR designs for Jaguar. TWR required a total annual budget of not less than £5 m, about the same as would have been required to field a team of British National Touring Cars ten years later.

Jaguar set aside an annual budget of just £1.3 m, whilst sponsorship broker Guy Edwards raised almost four times the Jaguar investment for the seasons, and that did not include America. This meant that the £4 m annual contribution from Silk Cut Cigarettes, and a further £1 m annually from

Castrol, covered seasons that included the vital 24 Hour Le Mans race. Edwards added a further $10m deal with British oil giant Castrol, one spread over three years for the seasons in which TWR contested American events.

Speaking from his base in Monaco early 2001, Guy Edwards cheerfully recalled: 'We must have wacked around £40m into Jaguar's racing over a six year period, approximated on the exchange rates at the time'. He also confirmed Roger Putnam's judgement that: 'Jaguar could not have gone racing at this level with TWR, without these significant sponsorship contributions to their budget'.

The Jaguar and Edwards sponsorship link continues into Formula 1, with Guy Edwards having brokered the Texaco/Caltex backing, as Technological Partners on the current R2 Grand Prix machines, an association that dates back to the original Stewart Grand Prix season of 1997. As Roger Putnam explained: 'For most of these seasons we were competing in 12 World Championship Group C events around the World, Le Mans and 16 IMSA events in the USA'.

He then added the following words that explain a great deal about the attitude of Jaguar towards the TWR run campaign, in the light of history:

'I believe we at Jaguar received the all-time motor racing bargain with these seasons. And it was such saleable success, of maximum value to the Jaguar brand, winning two Le Mans and three World Championships, never mind the American victories in the Daytona 24 hours, and other prestigious events, within our most important export market. Tom Walkinshaw and Guy Edwards are both outstanding businessmen and they received their rewards for being involved with us at Jaguar, but I strongly emphasise that Tom Walkinshaw and TWR did not make as much money as motorsport enthusiasts and the media popularly believe. In fact, because of TWR's slim profit margins, we changed the way we dealt with TWR during the later XJS seasons. Jaguar Cars continued to implement a different system of car ownership throughout the sports racing period.

'Normally we would expect to own the cars we raced, but in these later XJS seasons and throughout the sports car era, TWR retained ownership and could therefore sell the vehicles off at the close of each season to boost profits. Tom Walkinshaw did this very successfully with the XJS touring cars, and the XJR sports racing cars. My only regret, as Chairman of the Jaguar Daimler Heritage Trust, is that means we do not have some examples of our winning cars to display and use for sales promotion today. In particular I miss the presence of an XJR-14 in our collection, which dominated the World Group C championship in our final 1991 season.

'Overall I believe Jaguar can look back on that sports racing period with absolute pride, and my personal belief is that, as anticipated by Sir John Egan, motorsport added value to our share price!'

That concluded Roger Putnam's frank assessment, although he was not prepared to put a figure on the increase in the share value. However, outside financial sector advice would suggest that Ford paid an extra £2 a share (£8.50 was the successful bid) for the prestige motor racing had brought in the period leading up to their takeover of Jaguar Cars. Never has the added value of a specific motor racing campaign to the value of a brand been so publicly endorsed or calculated.

Exotic Jaguars and Customer Upsets

There is always a downside to such stories, and the problem that occurred with the Jaguar/TWR marketing campaign centred around the somewhat convoluted efforts of the two companies to build 'exotic car prestige', from their sports racing base. Unfortunately, the mass production based XJS coupes and XJ saloons described in Chapter 9, received far more customers than their rather costlier brethren. But behind the fact of that sales failure, there were some valuable branding lessons to be, perhaps painfully, learned here. Certainly, both Jaguar and TWR tried to exploit the link between the successful sports racing Jaguars and the showroom. However, the dialogue between Walkinshaw's Oxfordshire equipe and the Jaguar team at Coventry became muddled, and neither the alliance, nor the public, benefited from the ensuing confusion.

It is however important to be very clear that we are discussing a small number, actually only 335, somewhat exotic and expensive production cars. Machines that were destined for sale in the midst of a worldwide recession, that followed the boom years for the fastest Ferraris and Porsches of the late 1980s. Jaguar's challenge, in this 'multi-millionaires only' market, begun with the XJ220 of which 285 were built at Bloxham, now the site of all Aston Martin DB7 assembly. TWR had developed an entirely different machine, in a shorter time span, that was originally badged the V12 TWR 9R, later XJR-15, which was of more direct track pedigree, and rather more manageable size for a road car.

The XJR 15, unlike the twin turbo V6 XJ200 in production form, had the magic of the charismatic Jaguar V12 engine. Just 50 XJR 15s were built within TWR's Oxfordshire bases, and many were subsequently converted from track racers into fully road-legal machines at Kidlington, and subsequently shipped overseas. These rather exotic and yet road-legal machines, were particularly appealing to the ultra-wealthy Japanese. Not only were these machines exotic, let us also be very clear that we are talking about expensive Jaguars. The XJ220, cheapest at the official list price of £415 000, was at least double the price of an equivalent Ferrari or Porsche.

In any purchase of an XJR 15 from TWR, you would expect that more than £500 000 had changed hands, plus a charge of £80,000 if the customer wanted their '15' to become road-legal. The Jaguar racing cars that TWR retained in their ownership and sold on to private owners, realized from £400 000 for XJS coupes, to a regular £800 000 for the XJR ex-team sports racing cars, that is according to contemporary sources at TWR. Such was the current demand for racing history, that the cars which possessed that cachet commanded just over £1 m each, when sold with racing provenance of particular significance.

But the economic bubble was about to burst, and for many classic and supercar buyers it was a catastrophe, as it was for others. But when it did burst, Jaguar were left holding more than 100 unsold supercars, and an unpleasant prospect of taking legal action against potential XJ220 owners. These were wealthy individuals, or they had been before the crash came, who either could not pay, or would not pay, the final official price of £415 000. How Jaguar resolved this, and the story of the end of the JaguarSport alliance, is recalled below, but first a little about Jaguar and TWR's joint and separate exotic car projects in the 1985–94 period.

XJ220: A Protracted Birth

The prototype XJ220 was conceived amongst the 'Saturday Club' of off-duty Jaguar engineers, working with the then director of product engineering, Jim Randle. They would produce their vision of a contemporary Jaguar supercar, incorporating many sophisticated engineering innovations. These included a new 4×4 transmission system, a 6.2 litre variant of the V12 engine with 4 camshafts and 48 valves, in place of the production unit's 24 valves and 2 camshafts. The team also worked with the aluminium specialists Alcan, on a lightweight alloy monocoque, beneath an elegant, but elongated, 202 inch overall length body.

Encouraged by the team's recent motor racing successes, and with concern for the forthcoming sale of Jaguar, Sir John Egan decided to go public on this dramatic project. The sale of Jaguar was probably inevitable, given the political and economic climate of the late 1980s and Egan was concerned that the sale should at least generate considerable funds. At this point, however, it was the American automotive giant General Motors, rather than the Ford Motor Company, who were the favoured purchaser.

The new concept car was displayed at the 1988 British Motor Show, as the XJ220, the '220' referring to the expected maximum speed, in much the same way as the original XK120 had been named. The XJ220 attracted so much attention, that Jaguar asked TWR to investigate the practicalities of producing the car in sufficient numbers for a clientele apparently very keen to purchase, at considerable expense, rare and exotic automotive machinery. But Jaguar's ownership, the economic climate, and many other vital factors influencing the demand for such products was about to change, and the gestation period of the XJ220 was just long enough for the changes to hit the market, at the wrong time for the product.

The gestation period of the XJ220 from concept into limited production was from 1988 to 1992. The Ford Motor Company purchased Jaguar in 1989, and in December of that year Jaguar revealed a very different production version of the original XJ220 displayed at the 1988 Motor Show. Out went the bulky 4×4 transmission, the V12 engine, and the overall length was reduced by 11 inches (191 inches long), all of which reduced the kerb weight of the car by a total of 230 kg.

JaguarSport had built a new facility at Bloxham, outside Banbury, in Oxfordshire, with the expectation of making, appropriately enough, some 220 examples. At the height of the boom it was thought more likely that 350 would be made, as that was the apparent number of definite orders, each accompanied by a hefty deposit on the official retail price in the UK. This was not all that was to annoy prospective buyers; the original price rose dramatically, escalating from a provisional £290 000 to £415 000.

By the time production commenced at Bloxham late in 1991, the depth of the recession around the world was painfully apparent. Many loyal Jaguar customers paid the full price of the XJ220 and took delivery in 1992, but it obvious that a number of potential customers, who had already paid their deposits, were deeply unhappy. The Jaguar company was faced with the prospect of issuing writs to 140 customers who were refusing to pay the second instalments of their deposit of £50 000, due as the XJ220s entered final phase of completion.

No car company, but especially one owned by the second largest corporation in the world, is going to receive anything but adverse publicity from legal action against customers, particularly customers who were prepared to create a great deal of trouble. The production car did not comply with the original specification, and buyers were faced with a product whose value was lower because of the economic climate. Furthermore, there is no doubt that many of them had suffered from the impact of the recession.

So Jaguar had to decide either to take a large number of people to the Courts, or to compromise; Jaguar opted for a good old British compromise. Mr Howard Davies, the man who was then responsible for selling all of the 109 vehicles produced, and currently General Manager of Jaguar Daimler Heritage Trust, recalled in an interview in 2001: 'Rather than go to litigation against each potential customer separately, the last 109 XJ220s were brought back here to our Headquarters'.

Howard Davies, who had had the unenviable task of selling all of the 109 vehicles during the years between 1994, when production of the XJ220 had ceased and 1997 when the last 25 were sold, explained:

'we had storage for all 109 cars in a disused corner of the factory and a separate workshop for the XJ220 on our Coventry site. Those workshops remain here, oversubscribed with current XJ220 owners wanting factory service procedures enacted. I am glad to say we did sell all the cars, most at what I would describe as realistic prices. In fact, the last 25 went to Jaguar main dealers, Grange Motors of Brentwood Essex. Management eyes lit up when I took their cheque over for several million pounds!'

Once production of the XJ220 had ceased in 1994, the production line at Bloxham disappeared, and by the summer of that year had been transformed to become a Ford owned manufacturing site for the new, lower priced and higher volume, Aston Martin DB7. The new Aston Martin was the vision of the late Walter Hayes, who was convinced that, to be profitable, the company needed a higher volume car that would appeal to a wider (and younger) customer base; the story of Jaguar. The new Aston Martin DB7 design was the responsibility of the TWR company, who designed the car utilizing a Jaguar XJS platform and also manufactured its supercharged straight six engine.

So Jaguar had extricated themselves from a very tricky situation with dignity intact, and the TWR Group had benefited both ways. But there were some brand and marketing lessons to be learned from the public playing of the XJ220 saga, between 1989–97.

Supercar sales lessons – Part 1

1. The change in specification of the XJ220, from acclaimed motor show prototype to showroom reality, was too radical. Some customers cited the prototype's lack of an all-wheel drive (4×4) transmission, and especially the lack of a V12 engine, as sufficient reason to withdraw.

2. The confusing announcement in 1990 of the similarly expensive and fast XJR-15, which did have a V12 engine and a far more obvious racing pedigree, further reduced the desirability of the XJ220 in the view of some potential purchasers. See sub-section below, XJR-15, 'A Smaller Scale Success'.

3. The final £415 000 purchase price had escalated far beyond the written, 6 December 1989 pledge of '£290 000 plus taxes'.

4. Because of US legislation, there was no official American market for this Jaguar. Only when the final 109 were available, did unofficial means of individually importing the XJ220 to Jaguar's largest export market come to light. 'A significant number' of that final batch were shipped to North America by such means.

5. The six cylinder turbocharged XJ220 had no direct claim to TWR Jaguar's Le Mans and World Championship-winning V12 racing pedigree. It also used alloy monocoque chassis construction, whereas the race-winning TWR Jaguars and the XJR-15 used carbon fibre, state of the art racing materials technology.

6. The planned production numbers (350) were too high for true exclusivity. Again, the XJR-15's assured production maximum of fifty cars, was more attractive to the super-rich. Numbers can dilute exclusivity.

XJR-15, A Smaller Scale Success

Even as Jaguar were readying their prototype XJ220, for public display at the October 1988 motor show, TWR's Tom Walkinshaw, Andy Morrison and designer Peter Stevens, sat down at a meeting in Kidlington in August 1988, to 'discuss the project' that would become the XJR-15.

Work commenced in parallel to that on the XJ220 (Project 220 Ltd), by an entirely separate team from TWR Special Vehicle Operations, within TWR. At least by 1990, none of the parties involved was forthcoming on this subject, but it became obvious to Jaguar that TWR were developing another supercar, this time with Jaguar's legendary V12 at its heart. This resulted in some rather brisk discussions that ensued between TWR and the Jaguar management. With hindsight, this was the point at which the racing and road car alliance between TWR and Jaguar became strained.

TWR, under the R9R coding referring to the XJR-9 winning racing car on which it was based, had completed all the road legal endurance and legislative tests necessary by 1990. However, the project was unveiled to press and public in November 1990 as one aimed only at the most expensive of one-make racing series under the XJR-15 badge. Just 16 of the 6 litre, V12, TWR XJR-15s ran in only three motor races, although these high profile events supported the European Grand Prix circus, at the immensely fashionable and prestigious Monaco circuit, the British Grand Prix at Silverstone and Belgium's Grand Prix at Spa Francorchamps. The rewards were appropriate to the price of an XJR-15, and included an XJR-S for the

Monaco winner and a $1 m prize fund. The package was attractive to those customers looking for excitement and exclusivity.

It was probably for these reasons, that unlike the XJ220, the XJR-15 seemed to find enthusiastic buyers, and our contemporary information was that 43 out of the 'production run of 50 cars' had been sold by the November 1990 announcement date.

Supercar sales lessons – Part 2

1. Pedigree matters. Clients at this level demand a direct link to world class achievement and technology. The XJR-15 supplied both requirements.

2. Take a lesson from Ferrari. Make fewer vehicles than the order book predicts. In this rarefied atmosphere more can mean less.

3. Research who your customers are, who is going to buy the vehicle, before freezing specification and styling. TWR identified 30 positive prospective buyers 'round the world' before they committed manufacturing resources, and they limited production to fifty vehicles. Those original clients stayed on board, despite the same economic recession as faced by the Jaguar XJ220. Clients need to be as exclusive as the product.

4. Size does not matter, but function does. The XJ220 looked clumsy in comparison with the compact XJR-15. The XJR-15 had the primary functional features of the authentic Jaguar racing experience, transferred to the road.

The Ultimate Marketing Tool? Formula 1 Beckons

Although the majority of the expenditure from official Jaguar and outside sponsorship sources, was spent on sports car racing in the years from 1986 to 1991, the XJR coded V12 Jaguars actually appeared for the last time in January 1993, at the Daytona 24 hours. All three retired, and TWR elected to run a trio of XJ220Cs at Le Mans, in June of the same year. These three cars 'apparently' won the GT category, in a high profile entry backed by Unipart, and featured the 1997–2001 McLaren Formula 1 driver, David Coulthard. The word 'apparently' appears above, because the organizing club subsequently disqualified the TWR Le Mans entries, because they lacked catalytic converters.

This was the finale of the JaguarSport and TWR-Jaguar racing era, the successful partnership was truly over, and both parties would subsequently realize their Grand Prix ambitions, but by very separate routes. Tom Walkinshaw himself, rather than the TWR organization, became involved in engineering and even partial team ownership, with working spells at Ligier in France, now Alain Prost's eponymous Ferrari powered team, and with Benetton during Michael Schumacher's double World Championship era.

Tom Walkinshaw and TWR later purchased control of the struggling Arrows Formula 1 Team in 1996, which had then – and still has – to win a Grand Prix. The lack of Grand Prix wins does not appear to have dampened one Thomas Walkinshaw's determination to succeed, commitment to the Arrows Formula 1 Team or, to win. The Daily Mail's 'Rich Report 2001', ranked Mr Walkinshaw at 161st, on an equal basis with Mr Michael Jagger, with their fortunes estimated at £180 m. Reportedly, some £82 m of that amount was generated for Mr Walkinshaw from the sale of 75 per cent of Arrows equity in 1999.

Exactly how the revitalized Jaguar company and brand, now owned by and with the financial backing of the Ford Motor Company, realized some key marketing ambitions and became Formula 1 competitors, is the subject of the next chapter.

11

Achieving Global Brand Recognition

Ford Financial Muscle Paves The Way

Dr. Ing Wolfgang Reitzle, Chairman of Jaguar Cars when the Formula 1 project was launched in January 2000, made the keynote introductory speech of the Jaguar Formula 1 racing programme. Wolfgang Reitzle set out the company's agenda for a long term commitment to Formula 1 Grand Prix racing that, it is unofficially calculated, will cost Ford and Jaguar more than £680 m between 2000–2004.[1]

Best known as a senior member of top management at the highly successful German automobile manufacturer BMW, after a management career spanning nearly twenty three years, Reitzle resigned in February 1999, after the debacle following the crisis with Rover. Within a month, the Ford Motor Company had appointed Reitzle as Chairman of Ford's umbrella organization Premier Automobile Group in March 1999. PAG has specific responsibility for the Ford owned Jaguar, Aston Martin, Volvo, Land Rover and Lincoln brands. The American automotive giant had very clear ideas about the contribution that one of the key architects of BMW's brand development and maintenance over two decades, would be able to make to its own brand portfolio.

In his keynote speech, after recalling Jaguar's past glories outside Grand Prix, the message that Reitzle spelt out to more than a thousand crammed into makeshift grandstands at Lords that crisp January day began;

> 'our entry into Formula One, clearly signals Jaguar's direction for the future. There is an excitement and passion about Formula One racing which closely matches the emotional appeal of Jaguar. With its constant emphasis on advanced technology, Formula One racing will also be a valuable showcase, as we expand our model range and expand into new sections of the market, with innovative performance orientated products such as the X400 [subsequently named the X-type], our forthcoming new small sports saloon'.

---

[1] Costed from five years racing budget @ £340 m and three company acquisitions [£140 m] plus the estimated cost [£200 m] of a new corporate building to house all under one roof.

Throughout the year 2000, the theme of all Dr Reitzle's public pronounce-ments on Jaguar in Formula 1 was totally consistent, and always included the phrase 'Emotional Engineering', which Jaguar used alongside their mixed capital and lower case lettering for their ubiquitous advertising slogan, 'THE ART OF PERFORMANCE'. Dr. Reitzle explained Emotional Engineering as:

> 'that unique blend of emotion and technology which makes Jaguar one of the most emotive marques in the World. We call it emotional engineering, and the F type concept car which we unveiled at Detroit motor show two weeks ago, perfectly illustrates our philosophy for the future.'

Reitzle's Emotional Engineering theme encompassed more than establishing Jaguar's eagerness to take on the best of the best in the world premier motor racing league. Jaguar Cars are exploiting some of their Formula 1 R1/R2 designations, where the 'R' for racing prefix confers, for branding purposes, engineering excitement within a niche line of R-designated showroom Jaguars. This revives the Jaguar marketing tradition of linking competition glory with showroom appeal, that was recently evident in the TWR XJR prefixed Jaguar road cars. It is worth noting the similarity with the BMW tradition, in using M for Motorsport for a range of modified road cars that have sold 145 481 examples from 1978 to May 2000.

If Jaguar were to take their new products to a younger audience, they needed to compete and win against their main commercial rivals, BMW and Mercedes who are both front runners in this premier league of motorsport. Therefore, if Jaguar Cars Limited and their new, four-model line for 2001 are to go on the shopping lists of potential new owners, Formula 1 seems the most dramatic way of presenting the Jaguar name. Indeed, if the marque is to establish itself alongside the key showroom opposition, the use of Formula 1 Grand Prix, the FIA World Championship, appears the natural choice for their main marketing thrust. The TV and Internet audiences are measured by the billions, and have exactly the socio-economic profile that the company has identified for its new customers. (See Table 2.14 on TV F1 audiences).

However, such hindsight allows a clarity that was certainly not present in 1998, when the first moves in the direction of Formula 1 were made by American-based Ford top management, to key Jaguar executives. Roger Putnam, then Jaguar's Director for Sales and Marketing, recalled an unexpected item within a group telephone conference call. The call was relatively routine until the Jaguar management were surprised by the US based Ford Motor Company President and CEO, Jac Nasser, asking:

'If we made a big racing effort for Jaguar brand, putting it into Formula 1? Would that help push Jaguar forward?'

The reaction from the assembly of engineers, marketers and public relations executives listening to that statement, was immediately favourable.

'Then it all went quiet for a bit at the Ford end', recalls Roger Putnam. 'But there was never any doubt that we all thought it could be a good move to assist in repositioning Jaguar brand in our expectation of vastly increased sales.'

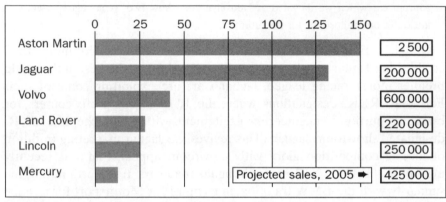

Figure 11.1   **Brand New Growth: Ford's Projected Sales Increase, 2000–05 (%)**
Source: Ford

As current Sales Director, with similar responsibilities stretching back to 1982, Roger Putnam was fully aware that Jaguar was about to go through the most enormous expansion. Growth that was designed to propel Jaguar Cars, from what had been a two-model line company at the time of Ford's acquisition of the company for £1.6 bn in 1989, with the XJS and XJ saloon, to a company with a five-model range. More importantly, that wider product range had to appeal to a range of customers far wider than the traditional Jaguar purchaser. In fact the same objective that had set the mind of Walter Hayes on the track of a new, less expensive but more attractive Aston Martin, thinking and action that led to the DB7.

Jaguar's new product range grew through the scheduled three model line of 1998, which comprised the XJ saloons, the XK8 which replaced the XJS coupe and convertibles, and the BMW 5 series and Mercedes E-class challenger, the Jaguar S-type.

Jaguar almost met their initial target, which was to double sales. Yet it had not been achieved through a simple injection of cash, as can be ascertained from Table 11.2 below. Initially, sales were buoyant; as the figures show, at the Ford takeover in 1989 the company sold 47 400 Jaguars. But both sides weathered some very tough conditions during the 1991–93 recession, and a number of internal manufacturing changes over difficulties and necessary investment obstacles did not help the situation. Then, Jaguar sales slumped between 1990 and 1992, from 42 754 in 1990 to 25 661 in 1991 and to a low of 22 478 in 1992, climbing modestly to 27 338 in 1993. These figures definitely tested Ford's faith in their expensive acquisition.

Recalling that worrying period in 1998[2], Nick Scheele commented: 'in 1992 the company had lost its direction and was losing money'.

In the same company house magazine feature, Birmingham based Ken Gibson continued: 'Scheele reveals that few people outside Jaguar reallsed how close one of the World's greatest car companies came to closing down in the early nineties'.

It was another five years, 1998 and a decade after Ford's acquisition, before sales exceeded 1989 levels, at 50 220. The advent of the S-type third line and the massive investment in facilities that took place in the 1990s totally, and very necessarily, transformed and overhauled the way all Jaguars were manufactured, and eventually paid off. There were improved sales figures from 1999, when 75 312 Jaguars were sold. In the year 2000, 90 031 Jaguars found new customers, almost meeting Ford's original expectation that the new S-type would double sales.

For global 1971 to 2000 sales statistics, see Table 11.2. For Jaguar's erratic sales in key export markets see Tables 11.2–6. All statistics were supplied by Jaguar Cars in January 2000.

The Future Branding and Sales Challenge

From the temporary and unassuming office of Jaguar Cars Ltd, Managing Director Jonathan Browning, the message for the future constantly goes out to local Jaguar showrooms, 'the next big hurdle on the way to a targeted 200 000 annual sales is production of the 2001 X-type. This will be a fourth major car line in Jaguar's fight back to worldwide recognition, the X-type is the smallest and most affordable Jaguar since the 1960's Mk2 range'.

Although not destined for large scale sales in 2004 and beyond, the F-type will benefit Jaguar with its E-type marketing associations and Jaguar's current Formula 1 presence, and the model's pedigree seems assured. The F-type will add a fifth line to the Jaguar range, but it is most important for its anticipated dynamic halo effect on the rest of the Jaguar range and the perceived appeal to younger potential buyers.

---

[2] Quoted in *Sovereign*, 'The Official International Magazine of Jaguar Cars', issue 24.

Table 11.2 Jaguar Cars Ltd: Annual Sales Figures Worldwide

Year	UK	USA	Canada	Europe	Japan	Rest of World	Total Export	Total Sales
1971	15 416							
1973	12 751	6 523						
1974	14 475	5 299	528	2 391		3 939*	12 157	26 632
1975	12 258	6 799	430	2 164		2 818*	12 211	24 469
1976	10 401	7 382	500	3 579		3 180*	14 641	25 042
1977	9 387	4 349	402	4 309		3 506*	12 566	21 953
1978	12 812	4 754	329	4 650		2 435*	12 168	24 980
1979	8 035	3 748	325	3 313		1 739*	9 125	17 160
1980	5 290	3 021	311	3 541		2 218*	9 091	15 011
1981	5 688	4 695	331	2 983		1 865*	9 874	15 562
1982	6 440	10 349	304	2 508		2 018*	15 179	21 619
1983	7 069	15 815	530	3 175		2 585*	22 106	29 175
1984	7 544	18 044	1 002	3 995		2 832*	25 873	33 147
1985	8 049	20 528	1 315	4 838		3 015*	29 696	37 745
1986	7 579	24 464	2 032	4 332		2 564*	33 392	40 971
1987	11 102	22 919	2 660	6 550		3 412*	35 541	46 643
1988	14 504	20 727	2 154	7 876		4 233*	34 990	49 495
1989	14 243	18 967	1 606	8 199	1 856	2 529	33 157	47 400
1990	10 664	18 728	1 005	8 094	2 502	1 761	32 090	42 754
1991	5 809	9 376	740	6 304	2 438	994	19 852	25 661
1992	5 607	8 681	581	5 021	1 501	1 051	16 835	22 478
1993	6 215	12 734	630	4 866	1 510	1 383	21 123	27 338
1994	6 685	15 195	713	4 641	1 445	1 341	23 335	30 020
1995	8 798	18 085	945	7 215	2 311	2 373	30 929	39 727
1996	8 422	17 878	1 051	6 917	2 335	2 398	30 579	39 001
1997	9 524	19 514	1 020	8 665	2 366	2 686#	34 251	43 775
1998	11 670	22 503	1 407	10 522	2 009	2 109#	38 550	50 220
1999	15 483	35 039	1 700	17 015	2 367	3 708#	59 829	75 312
2000	15 003	43 728	2 315	21 401	2 880	4 704#	75 028	90 031

KEY: *Including Japan # Excluding Japan.

Table 11.3 **Regional Sales: Number of Jaguars Sold Between 1970 and 2000 in Five Key Markets**

Years	France	Japan	Italy	Germany	Spain
1970 (1 Oct 69 to 30 Sept 70)	906	172	622	425	62
1971 (1 Oct 70 to 2nd Oct 71)	996	193	669	982	60
1972 (1 Oct 71 to 30 Sept 72)	551	97	143	253	22
1973 (1 Oct 72 to 30 Sept 73)	512	170	507	408	213
1974 (1 Oct 93 to 30 Sept 74)	565	224	409	186	71
1975 (1 Oct 74 to 30 Sept 75)	760	363	350	543	63
1976 (Oct 75 to Dec 76)	1051	406	597	1983	72
1977	849	501	475	1589	44
1978	453	531	365	2011	56
1979	400	335	279	508	13
1980	418	908	695	1527	63
1981	206	90	676	2510	36
1982	201	233	439	845	81
1983	260	304	397	1220	91
1984	262	379	283	1938	86
1985	406	371	332	2350	115
1986	512	464	264	1852	238
1987	1026	588	610	2184	405
1988	1515	1124	804	2407	516
1989	2018	1868	850	2410	519
1990	1906	2502	883	2479	364
1991	1200	2438	835	2146	334
1992	750	1501	644	1881	223
1993	600	1510	422	2014	225
1994	605	1445	493	1430	418
1995	717	2311	1081	2525	633
1996	594	2335	1002	2512	490
1997	802	2366	1050	3151	728
1998	860	2009	1325	4059	781
1999	1550	2367	2206	6090	1710
2000	2302	2880	3797	6467	2157

Instead of the publicly quoted 200 000 total sales volume, when the X-type is established, senior Jaguar and Ford employees know that the possibilities are greater. If the sales messages can be truly successfully implanted, with a new generation of buyers heading into Jaguar showrooms, the Halewood Escort factory that Ford converted for X-type output, could produce up to 300 000 X-types annually[3]. That would mean Jaguar occupying the 400 000 annual output level by 2004, becoming the largest scale car brand 'Made in Britain'.

If any corporation knows the values of the economies of scale that can be achieved through mass production, it must be Ford. Moreover, Ford executives are well aware that even at annual sales levels of 400 000, Jaguar output would be less than the half-year deliveries attained by Mercedes or BMW in 2000. Mercedes sales are at 'just over one million for the first time', said a UK company spokesman in February 2001, and the BMW company sales level is at 822 180. Even the comparatively small Audi division of Volkswagen Group delivered 653 404 cars in 2000, versus 90 031 of Jaguar. There is still a large gap to close before Jaguar even nibbles at the heels of the smallest German prestige car manufacturer, so Grand Prix truly represents Ford and Jaguar's grand global ambition.

Although, historically, Jaguar's motorsports heritage never included participation in the Grand Prix arena, both the Ford and Jaguar management were convinced that a new approach was needed, in order to revitalize and widen Jaguar's appeal. Currently, because of the transformation wrought on the sport and spectacle, Formula 1 represents the world's greatest marketing opportunity, certainly as far as the automotive sector is concerned. The company's planned sales expansion could only be achieved through a marketing programme that was linked to the world's widest audience, that would also equally expand Jaguar's sales prospecting.

Whilst recent research has failed to reveal that Jaguar apparently undertook any heavyweight market analysis, or had a succession of lengthy meetings to debate their move into Formula 1, the decision was taken. A £140+ m Ford investment in sporting hardware (detailed company acquisition costs follow), and a reported[4] annual budget of £85 m per annum, were all demanded to sustain a presence for Jaguar in the 2000 to 2001 Formula 1 season. That is not including the estimated £200 m price for new premises required to accomodate all the racing divisions under one roof. A project now delayed, and more of that later.

---

[3] *Sunday Times*, 4 February 2001.
[4] *Autocar*, 10 January 2001.

These somewhat heavyweight decisions were apparently, all enacted without the usual multinational, corporate layers of the decision process. An ardent motorsport supporter since his working life began in Ford Australia, Jac Nasser instigated what was needed for Ford to switch from the blue oval branding in Formula 1, to Jaguar. That appears straightforward enough, given that Ford had been associated with powering Grand Prix cars since 1967, through Cosworth Engineering in Northampton. In fact more than the £140 m quoted above was required, because Ford was changing both its relationship and branding in Grand Prix.

Ford had to move almost as fast as one of their earlier Stewart Racing GP cars, to advance its plans for Jaguar and to defend its earlier investments in Stewart Grand Prix. In September 1998, Ford paid £50 m for Cosworth Racing[5]. In 2001 Cosworth Racing had a Ford reported turnover of £112 m, and employed 725 people in Northampton UK, and Torrance, CA, USA. Much of Cosworth's motorsports business always came from Ford, split into three primary areas during 2000–2001:

1. Design and supply of CR-prefixed racing V10 engines, first to Stewart Racing from 1997 to 1999, then Jaguar Cars, 2000 onward.
2. Design and supply of the turbocharged X-family V8 engines, to CART (Championship Auto Racing Teams), America's premier single seater formula, since the 1970s to date.
3. Conversion and modification of turbocharged Zetec Ford 4-cylinder family engines, for the Ford Focus World Rally Championship programmes of 1997 to date.

Former free marketers who sold to the world motor industry on three primary fronts, racing, patented castings and limited specialized car production, Cosworth, *post* the Audi and Ford takeovers, became a fragmented business. It is totally relevant to our Jaguar Formula 1 story, as it is destined (as is the Pi Group Ltd detailed below) to be housed with Jaguar Racing at a new HQ.

These leading edge premises would have demanded an investment (as far as is possible to ascertain), of more than £200 m from Ford Motor Company/Jaguar Cars, a figure which is comparable to the £170 m quoted for the Mclaren TAG, Paragon HQ, due for completion 2001 or 2002. The new Silverstone complex was planned to accommodate Jaguar Grand Prix, Cosworth Racing and the Pi company, grouping the alliance together on a massive site, housing more than 1100 employees. It would appear that, because of planning issues and negotiations with the BRDC (the BRDC and

---

[5] *Britain's Winning Formula* (Macmillan Business, 2000).

Jaguar were reputedly £200 000 apart) no firm completion date had been announced for these combined Jaguar Racing, Cosworth and Pi premises. On such small grounds it would appear that the whole project has been placed on hold. It is now unlikely that the trio of Ford racing divisions will move in to the Silverstone complex within the planned 2003 to 2004 schedule, if at all.

Hindsight may suggest that the purchase of Cosworth Racing was a manoeuvre, part of a logical Ford corporate plan, to acquire their regular racing and rallying engine supplier. That would not be the complete truth, as Ford was offered ownership of the Cosworth company routinely, in their long 40-year association (from 1959 to date), but previously they had always declined. Now, Ford faced an urgent need to purchase Cosworth, because the complexities of multinational corporate acquisition had left the Volkswagen Group's Audi AG, the owners of three Cosworth divisions, in the aftermath of Volkswagen's successful bid for Bentley at Crewe. The deal was done for the Cosworth Racing Division as fast as possible. By September 1998 Ford secured what would be Jaguar's Formula 1 engine supplier.

So far as the Jaguar project was concerned, Ford was going to have to undertake to buy more corporate assets, but this time on a planned schedule. In June 1999, Ford announced the successful completion of its plan to purchase Stewart Grand Prix, a company they had initially partly financed, and that company's £5 m premises at Milton Keynes. As ever at Ford, no price was officially confirmed, but in the Guardian of 13 December 1999, it was reported that the Stewart team was 'eventually sold', at an 'estimated £60 m'. The *Sunday Times* in June 2000 went further, and reported 'Ford paid around £70 m'. Since Ford had initially provided substantial funding for the Stewart, they could have hoped they would have paid less to acquire the company. In fact they had invested a contracted £60 m during the three years from 1996, the £20 m a year including the costs of supplying Cosworth's engines. In spite of their early commitment, it is unlikely that they were under any illusion as to the price they paid.

It is also important to emphasize that Ford's initial act of faith, confirmed in December 1995, was crucial in establishing the embryonic team's credibility. The Ford name, and money, had certainly eased the path for Jackie Stewart and his partner son, Paul Stewart, in attracting the balance of annual operating costs (then a reported £25 m per annum), from major sponsors such as the HSBC and the Malaysian Tourist Board. Stewart Grand Prix had contested the 1997, 1998 and 1999 seasons with initial heartbreaks, and subsequently remarkable success. The 1999 team that Stewart sold on to Ford and Jaguar, had won one Grand Prix, been placed in the top three on occasion, and was set for more achievement. Some GP teams have not matched the Stewart GP record in over 15 years' existence on the grid.

But Ford still had more to pay for the necessary resources they would require to make their vision of transforming Jaguar's Brand and position (to say nothing of the spin-offs onto the Ford Brand itself) through motorsport. More accurately, the technologies now required to achieve any front-running position in motorsport includes new materials, technology, information, aerodynamics, engine and other engineering skills. Ford had to, and did, acquire a mixture of tangible electronics and leading-edge software from their purchase of Pi Group, the Cambridge based employer of 275 in the UK and USA, in November 1999. As dictated by usual Ford practice, no takeover price was released, but we know that Pi Group had a turnover of £12 m in 1998, and a high growth rate which forecasters interpreted as a turnover of £30 m by February 2001. It seems likely that Pi's takeover price was less than £20 m.

Like so many managerial decisions and acquisitions in the 1996–2001 Ford/Jaguar Formula 1 brand management revolution, this move was implemented and managed by Neil Ressler. An American charged by Ford with responsibility for racing programmes and high technology related areas, Mr Ressler 'elected to retire' from his final position as Chairman of Jaguar Racing in February 2001. Over a vital period of five years, Mr Ressler's understated style had implemented the grand Stewart Ford and Jaguar Racing Formula 1 plans, including the year restructuring of Jaguar Racing in 2000, against a background of disappointing race track results.

Neil Ressler left Ford when former triple Formula 1 World Champion driver Niki Lauda, was appointed Chief Executive Officer of Premier Performance Division, also in February 2001. This new division was the motorsport subsidiary of Ford's international Premier Automotive Group (PAG). Having sold Lauda Air to Austrian Airlines, Niki Lauda expected to be 'back in racing on a full time basis'. Another former racer, Bobby Rahal of the USA, was appointed as Chairman of Jaguar Racing in 2001. Rahal has led Jaguar's racing management since then, whilst Lauda has co-ordinated the Jaguar, Cosworth and Pi company's individual efforts in the common cause, prior to their, now postponed, move into a purpose-built set of premises at Silverstone.

After years of being accused by purists within Formula 1 circles of lacking commitment to racing in this category, Ford are spending extremely serious money on racing two 'Jaguars' that underperformed as at mid-2001. In spite of their early promise racing under the Stewart name, the Jaguar team had scored just four championship points during the 2000 season, finishing at the tail end of the final FIA 2000 Constructors' rankings. However, those rare points were taken at the highest profile track, the Monaco street circuit, where Jaguar were also third in 2001.

Formula 1: The Cost of Extending the Brand

The recent announcement that the whole of Ford's Formula 1 enterprise will not be under one roof at a new facility at Silverstone, at least not by 2003, has put back the company's plans. It is always difficult to predict such events, but current difficulties with the local planning authorities have certainly delayed the future integration of the three companies, Jaguar Racing, Cosworth and Pi Group. Whether this will reduce the total expenditure of the Ford Motor Company, estimated at a minimum of £680m by 2004, is another matter. The actual costs of the Silverstone facility, estimated at some £200m, will probably not be saved in total, as it is likely that an alternative solution will be found.

All this has been based on the determination and vision to expand Jaguar brand appeal, through Grand Prix hardware and four seasons of F1 racing budgets, amongst other marketing programmes. These figures are dwarfed by the £1.6bn purchase price of the Jaguar brand, but they are the biggest investments these authors have ever encountered in current motorsports.

Indeed, it was the sheer size of the financial and engineering commitment, that gave many who observe Ford business practice closely, pause for thought. Externally, the insertion of the Jaguar brand instead of Ford, for the Formula 1 activity, looked simple. In fact, immediate and longer term decisions, and substantial financial investment, had to be made almost simultaneously. This was because the price of 21st Century, Grand Prix participation with a team that cannot run at the back of the grid, has escalated. Perhaps the recent furious activity by the group of major manufacturers currently involved in the sport, indicates their determination to have a greater say in the sport's structure; they want their money to talk! The debate between Bernie Ecclestone's company SLEC, Kirch the German Television Group and Fiat, BMW, Mercedes, Ford, Renault and Honda will no doubt be resolved, but not without the big six protecting their continued investment.

Whilst the Ford Motor Company have never publicly revealed what they spend on motorsports at any level, those with calculators handy can check how some of the routine costs mount up. In 'Jaguar 2000', a recent press release, the following primary consumable costs were published. We have added the approximate numbers required in a season, and multiplied by the individual price for each item, to calculate a season's total.

	Total £
70 engine rebuilds @ £180 000 each.	12 600 000
7 monocoque chassis £90 000 each (build cost).	630 000
15 suspension sets @ £ 35 000 each.	525 000

	Total £
12 steering wheels with electronics @£50 000 each.	600 000
10 driver seats @2500 each.	250 000
40 exhaust sets @ £6000 each.	240 000
200 race wheels @ £750 each.	150 000
14 front nose wing assemblies @ £3000 each.	42 000
34 sticker sets @ £1000 each.	34 000

And so the £14.72 m weekend and mid week practice shopping list grows, but that is only a fraction of what is actually spent each season. That £14.7 m is just a sample of routine costs, as it takes no account of running three separately housed companies, on a combined payroll of 1360 in 2001. Now it seems unlikely that they will be housed under one roof, at least not on the current information! The team are currently paying their leading driver, a reported £21 m over three years[6], travelling to 17 race circuits and many more practice sessions.

What of the other major teams? Here we enter the imprecise world of estimates. The following data has been calculated after interviews at director level, inside and outside Ford owned divisions, plus discussions with long term media and Formula 1 related employees. Overall, the information received from these sources, revealed a remarkable consensus regarding the league table of spending in the 2000 Formula 1 season, and probable 2001 budgets, for leading teams. It is unlikely to be purely coincidence, that in the case of three of the four teams mentioned (the exception was Jaguar) their spending rates coincided with their finishing order in the 2000, FIA Formula 1, World Championship for Manufacturers.

Out on their own, with an annual budget of over £100 m, were Fiat-backed and partially owned Ferrari, who were alongside the West Cigarettes-supported McLaren Mercedes equipe. Over some 30 years the Fiat parent and their Ferrari offspring have also benefited from huge cigarette sponsorship revenues, from Marlboro. It is important to note that Daimler Chrysler have a 40 per cent holding in the TAG McLaren Group, the umbrella for five Mclaren companies. The balance of ownership is with TAG and TAG McLaren Chairman and CEO Ron Dennis, who each hold 30 per cent of the equity.

Also spending similar budgets in these estimates, is Ford with Jaguar, placed alongside BMW AG, with Williams Grand Prix Engineering. Published quotes here centre on expenditure of £85 m each in 2000, but spent on very different approaches to the business of extracting maximum commercial advantage from Formula 1.

---

[6] *Sunday Times*, 25 June 2000.

Marketing and Public Relations Strategies: BMW v Jaguar

Jaguar initially spent their marketing money at a huge rate in early 2000, 'painting the town green' in the words of one Jaguar director. BMW concentrated more on a conservative in-house engineering presence that also suited the traditional 'let the racing do the talking' approach of Williams' directors.

Roger Putnam recalled the Jaguar marketing and branding priorities in 2000:

'Our job was to immediately put Jaguar's presence up alongside that of Mercedes, who were established former champions in Formula One, and BMW. We had to give Jaguar real international presence and establish that we were no longer interested in just being niche players. We are not going to reach new customers and expand sales to toward the quarter million mark by being shy!'

To achieve this visibility alongside the bigger global brands, Jaguar targeted the Australian opening round in Melbourne, for the corporate green treatment and the mud-ravaged British Grand Prix, ran on a untraditional, early Spring date in 2000. The Australian event certainly saw the Jaguar brand make an impact, from painted trolley buses onward. But at the British Grand Prix, British headlines, outside the specialist press, were harder to dominate, as the Silverstone organisers had to close the car parks on Saturday's official practice day, because of the seas of mud. Whatever your marketing plan, it's worth remembering that the British weather can undo even the most carefully laid brand assault!

Unfortunately for Jaguar's corporate pride, BMW/Williams Grand Prix were runners up to Ferrari (170 points) and McLaren (152) in the 2000 Constructors Championship on 36 points, whereas Jaguar Racing took ninth and last place in their debut season, with just 4 points. Jaguar's number one driver Eddie Irvine, a contender for the 1999 World Championship driving a Ferrari, scored those points and that placed him 13th of 15 points-scoring drivers in the FIA 2000 Formula 1 Drivers Championship.

Although Jaguar had dominated the media headlines at their launch in January 2000, there was inevitably some adverse media reaction to the season's setbacks. These criticisms were not just confined to the specialist British press. The influential *Sunday Times* correspondent Peter Windsor noted:

'Ford bought Stewart last year when that team was on the rise, in 1999 Stewart scored a win, a second and two pole positions. With a larger budget, Ford could at least have expected to follow on from that.'

Money does not always buy success in Formula 1. Indisputably, Jaguar did badly on the track, but Windsor also took issue with the massive marketing effort, acidly commenting:

'it has become very tacky in terms of image. For example, an horrendous chrome "leaper", as the ad men wanted to call it, adorning an ostentatious motor home at the British Grand Prix,'

said the disappointed *Sunday Times*, Formula 1 correspondent in the Summer of 2000.

Another painful comparison for Jaguar and Ford, not just on a similar expenditure basis, was that with BMW in the same season's Formula 1 circus. BMW's winning record from the early eighties raised German expectations in 2000, so BMW's Munich-based marketing and PR personnel denied any suggestions of front-running expectations, in every public proclamation. That PR strategy worked well with the racing insiders, and knowledgeable spectators of both Britain and Germany, but Jaguar's management, particularly PAG Chairman Wolfgang Reitzle, then stated that the Jaguar brand obtained a commercial advantage from merely participating in Formula 1, even without winning form.

From a marketing viewpoint, the difference in style between BMW and Jaguar has closed during the 2001 season. Jaguar have toned down their PR launch, from a 1000 person party at Lords Cricket Ground in 2000, to a more workmanlike 2001 presentation for less than half that number at their Whitley, Coventry Engineering Centre.

There were however, bigger marketing battles that Jaguar did win. The company were winners in the virtual reality world, for 'Jaguar Racing.com' was named 'Best World-Wide Sports Website' at the 2000 Emma Foundation Awards in November 2000. Jaguar finished runners, up alongside globally established www.ft.com, in the Best News category. In eleven months from its establishment on 25 January 2000, 'Jaguar-Racing.com', has recorded a massive 1.772 million visitors, 91 million hits and attracted 67 000 Virtual Club members.

Hardcore racing appeal at the Jaguar site was not so strong, just 2600 enrolling as Jaguar Racing Club members. The company also dispensed with their six issue 'Youth Appeal' glossy racing magazine Jaguar Racing, in 2001, replacing it with a more conventional quarterly house magazine, simply titled Jaguar, that incorporated racing content. These facts support Dr. Reitzle's theory, that participation was the most important element for Jaguar in their first Formula 1 season. There is also some reason for Jaguar's marketing and PR strategists to hope that Jaguar will eventually come second only to Ferrari in its appeal to racing fans. Such success, and therefore sales of associated clothing and accessories, would be valuable, judging by public support in 2000, especially in the USA Grandstands of Indianapolis.

However, Dr. Reitzle did admit at his introductory speech in January 2000, for the Jaguar R1 model: 'our ultimate objective is the ultimate prize in motorsport, the FIA Formula 1 World Championship.'

There is little doubt that we can expect this enormously costly hunt for Jaguar Grand Prix glory to continue, with associated fascinating marketing initiatives. Formula 1 for Jaguar, will last just as long as Ford feels Jaguar can not just dent the German opposition, but also establish a new world benchmark in profitable prestige car sales for generations to come.

PART IV

CONCLUSIONS

12

Lessons from the Campaigns: The Successful Branding Process

A Summary of the Two Campaigns

These are two campaigns that were, and are, rooted in the fundamental marketing philosophy that Brand Development can be effectively and efficiently achieved, at least for an automotive brand, through the utilization of motorsport. However, the two campaigns were fundamentally different in their detailed objectives and in the manner in which they were established and managed. The Jaguar programme that we have examined had extensive historical roots that go back to the early days of the company's history and continued sporadically, through the rather darker years of commercial disaster under the control of British Leyland and into the present Ford ownership, the latter company taking Jaguar into Formula 1 for the first time in its history. Although the Jaguar brand managed to retain some elements of a sporting legend in the 1970s and early 1980s, poor build quality and un-reliability, that was the antithesis of the original brand essence, had severely tarnished the company's products by the 1990s. Alternatively, the campaign undertaken by Fuji Heavy Industries' car manufacturer Subaru was based upon a very much more restricted and, as far as their main target markets were concerned, almost complete absence of a sporting history, but the product was the essence of the brand.

In that sense, the two campaigns not only started from different positions; they were formulated upon fundamentally different points of departure as far as the marketing of the products were concerned. The more pragmatic approach traditionally associated with Jaguar's motorsport activities, contrasts starkly with the persistent and consistent efforts made by Subaru through the World Rally Championship. The Jaguar brand was in a position of needing resuscitation and revitalization, whereas the Subaru brand had no reputation in the major markets of Europe or North America. In addition, although the time span of the two campaigns was similar, approximately a decade in each case, they took place over different periods, from 1984 to 1994 in the case of the TWR and Jaguar partnership and from 1990 to today in the case of Prodrive and Subaru.

157

There were a number of other areas of difference, which were probably more important than simply time. The Prodrive/Subaru partnership was founded upon very clear and carefully analysed principles (see Chapter 3) plus a very strong commitment to achieve success. Furthermore, Subaru were very clearly aware of the importance of the 'softer' aspects of managing the partnership. In this context the engineers and the managers, particularly the leader of the Subaru team Mr Takemasa Yamada, had identified the specific behavioural principles that still drive the partnership today (see Chapter 3). The Jaguar campaign involving TWR was directly targeted at a single event, the 24 Hours Le Mans race and, as with the Subaru campaign, Jaguar relied very heavily upon the technical expertise of the TWR racing organization to achieve that success. The extraordinarily heavy emphasis that this campaign placed upon the popularity and marketing advantage of a single event, the 24 Hours Le Mans race, was expressly different from the long term campaign that Subaru waged in the WRC. These factors, plus the consistency and determination of clear marketing objectives which were applied in the case of Subaru and Prodrive, are distinctly different from the experience between Jaguar and the TWR Group.

In the case of Subaru, the brand was relatively unknown outside of the Japanese market and, as far as the majority of customers and potential purchasers were aware, the company's products were essentially utility vehicles for use on farms. The company's World Rally Championship campaign was their first venture into a prolonged motorsport marketing effort, although it had been preceded by an abortive attempt to enter the Formula 1 arena (see Chapter 3, p. 51). As is clear from the sales figures detailed in Table 5.1, p. 78, from 1993 the company's export sales of the Legacy, Impreza and Forester models rose steadily, both in total numbers and as a proportion of the company's total sales. The fact that the company's products became steadily better known and linked to the essential qualities of the successful rally team, were an important and vital element of the product's surging sales, particularly in major export markets.

Jaguar had campaigned in the sports racing arena for more than fifty years, on a rather sporadic basis. Their racing heritage and image was of far greater length and had been imposed on a number of levels, from saloon car racing, through sports cars and then finally, after the decade with TWR, to Formula 1 under Ford's ownership. In the first part of the modern campaign, at the beginning of Jaguar's somewhat disastrous period under the management of British Leyland, the management were persuaded to undertake a sports car campaign. The decision was taken to use the successful Broadspeed company to develop a racing car out of the 12 cylinder engine and with the car based on Jaguar origins. The resulting campaign over two seasons was disappointing and lacklustre, the cars racing eight times without a win, often failing to finish. In addition the sales and marketing managers of British Leyland were often somewhat careless and jingoistic in their press statements, which often led to the raising of expectations that were unfulfilled.

Jaguar's successful partnership with TWR, first, when the company was still part of British Leyland and then, after privatization in the 1980s under the Chairmanship of John Egan, was a vital element in developing the company's stronger position and brand strength at the time. Egan was determined to exploit the brand, and the new Jaguar management had a clearly focused approach and was very determined to prepare the company for regeneration and build upon their previous success. In that respect motorsport was to play a crucial role and, from the evidence, a role that was exceptionally cost effective in achieving the Board's objectives, particularly in relation to the price Ford paid to acquire the company (see Chapter 10).

However, as in the past thirty years, both under the company's founder and during the 1980s and 1990s, Jaguar continued to be a somewhat peripatetic participant in motorsport, without any long-term strategy. As we have chronicled in earlier chapters, initially the company's approach was almost on an event by event basis, and certainly only considered when sales of vehicles would be the end result. The early Le Mans entries and the later sorties into American events were all specific and individual, or at least covered a narrow time span. Also, the company tended to enter whatever events seemed at the time to be appropriate, from touring cars to the exotic heights of world sports car racing. Furthermore, they also approached all but the least expensive racing activities with the assumption that they would fund at least some of the expense through the pockets of sponsors. Whilst the latter characteristic is hardly exceptional, particularly since 1980, Jaguar's approach does contrast very significantly with that of Fuji Heavy Industries and Subaru.

The approach taken by the Japanese company was therefore very different from that of the Jaguar campaign. The nature of the differences are illuminating in the context of understanding both the process of building and maintaining brands as well as the theory of branding. The development of such campaigns are a vital element, although only one option, in the overall development of a brand's attractiveness, particularly in relation to brands that offer, or are attempting to offer:

'distinctive specialist products or services at premium prices, to a population of customers with highly differentiated needs'.

In both cases the primary objectives of the campaigns were to clearly establish the individual brands as identifying distinctive specialist products, available for premium prices and able to satisfy discerning customers with differentiated needs. It was clearly understood by both companies that such campaigns do not, of themselves, sell individual models; the campaigns identify a set of clearly defined criteria that satisfy the differentiated needs of the group of target customers.

In both cases the companies had accepted that they would never be able to create the volume of sales that would achieve economies of scale to enable them to compete with the major manufacturers. In addition, Subaru and Jaguar had identified the specific nature of the product characteristics that they wanted to deliver, and both were somewhat similar. In Jaguar's case it was 'PACE, GRACE and SPACE' and in the case of Subaru it was 'SPEED, STYLE, SPACE' (p. 70).

In the same way both companies had concluded that with over-capacity in the industry, mature markets and the changing demographics of the societies in all the main markets, to attempt to run with the mass producers would be suicide. Their conclusion was to pursue the consumers who were prepared to pay a premium price for a premium product. But, how to convince consumers that the products are premium, and that the products have the desired characteristics? In this case motorsport is an avenue that can produce the desired 'halo' effect on the brand and therefore place the product in the desired segment.

However, the fact that the two companies had similar bases for their actions did not mean their approaches were the same, since they started from very different situations. Jaguar already possessed a sporting reputation and whilst the products had denied customers the brand promise, the image was in place. Subaru had no image and therefore no product promise, it had to be established. Furthermore, in terms of the approach taken by each company, it is evident from our research that there are the following distinct differences.

Planning and Preparation

Subaru: Fuji Heavy Industries had already engaged in an expensive and abortive attempt to enter Formula 1 (see Chapter 3), but that experience did not deter them from motorsport. However, their approach to entering the World Rally Championship was systematic, carefully planned and cautious. As we record in Chapter 3 the company were exceptionally clear about their aims and objectives, to the extent that they had clearly defined them before they began to establish the beginnings of their partnership with Prodrive (see Chapter 3, pp. 54–55). Mr Kuze and Mr Narita (see Chapter 3, p. 52) were also convinced, and persuaded the FHI Board in 1988 that for any marketing initiative through motorsport to be successful, it was necessary for it to be a sustained campaign, not a 'one-off event'. Their objectives in entering the WRC were set out in 1988, as follows (see also Chapter 3, pp. 54–5);

'THE AIMS AND TASK IN WRC COMPETITION
To prove the excellent basic engine layout in driving performance
To show continuous challenge to appeal to the public with Winning.'

Source STI

- To provide the company with an advertising opportunity to place the characteristics of the company's products before current and potential owners.
- To provide an additional opportunity and platform for the technical development of the company's products.
- To differentiate the products through distinctive branding and imaging in domestic and export markets. Subaru participated in the Group N rally series to give the cars homologation.

Jaguar: Jaguar acted independently in the first stages of their motorsport activities under the leadership of William Lyons, and only latterly used partners, both in North America and Britain to undertake their racing activities. It is clear that at each phase of the motorsport efforts the company always had limited and specific time frames in which to achieve their stated goals. Under Sir William the priority was to achieve VFM, Value for Money which equated to always ensuring that there were no 'open-ended commitments' in their racing programmes (see Chapters 10 and 11). Furthermore, the company has always appeared to have a pragmatic approach to the aims and objectives, usually achieving specific goals (winning at Le Mans), and then withdrawing.

Partnership Selection

Subaru: In their initial investigations of the marketing opportunities offered by the World Rally Championship, Fuji Heavy Industries were clearly concerned about the somewhat bruising, and expensive, experience they had had in their Formula 1 journey. They were also of the view that single motorsport events were of limited value in marketing (Chapter 3, p. 52): 'a sales person cannot talk about one event in 1989, forever!' Having decided on their overall strategy and the desirability of the WRC, the company's search for an appropriate partner was careful and systematic. In addition the company's manager selected to lead the effort spent a year with Prodrive, before the final decision to partner was taken, and to take a long-term view of the campaign.

Jaguar: In their early partnership with TWR, Jaguar had limited aims and objectives that certainly evolved over time, exactly as the Prodrive partnership has with FHI. However, at the outset the aims were less well defined and certainly, in the latter parts of the World SportsCar Championship campaign with TWR, it became clear that the two companies were 'not precisely in tune', particularly with respect to road car derivations (Chapter 8, p. 110). The statement of how the Ford parent later took the company into Formula 1 (See Chapter 11, p. 140) is an indication of the different approaches. It is important to point out that Ford had already financed the Stewart Grand Prix team and were rather more familiar with the benefits to be gained from motorsport, in spite of, or perhaps because of, their somewhat peripatetic involvement. However, in this context the approach of the two organizations to partnership selection was, including the decision by Ford to purchase Stewart Grand Prix, somewhat different.

The Need for Differentiation

In both cases the general need for the two companies to use the medium of motorsport to create differentiation and brand identity was clear, as is detailed in Chapter 1. The manner that each of these companies chose to implement their campaigns, and precisely what differentiation they set out to achieve, was distinctive.

Subaru: In the case of Subaru, Fuji Heavy Industries were in effect creating a brand, not redefining or resurrecting one. The company had, however, already "bet its future" (See Chapter 3, p. 54) in terms of developing a new range of engines and models, more appropriate to the European and North American markets. Therefore, the need for them to differentiate their products was crucial and they had to create a new customer proposition in terms of the characteristics of the brand. The phrase 'inexpensive and built to stay that way' may have been attractive in the Japanese agricultural market, it would not sway the sated customers of the two largest car markets in the world.

The company had clearly set out the characteristics of their preferred customer and the same report (STI 1984) also identified the essential characteristics of new products:

'Instinctively aggressive and competitive and wanting a product that appeals to

their instinct.

'The cars offer SPEED, STYLE, SPACE.

Source STI

There is an interesting parallel between the characteristics of the Subaru range and the Jaguar phrase 'Grace, Pace and Space'.

Jaguar: In the case of Jaguar, for Ford there was clearly a need to expand the potential customer base, in exactly the same way that Walter Hayes had determined that the appeal of Aston Martin had to be expanded. In the case of both companies they set out to create new models with the same essential product characteristics, and the Aston Martin DB7 and the new Jaguars, are designed to achieve that aim. Therefore, whilst the companies did take the same route for essentially the same reasons, the time it has taken for them to reach their goals has differed, as has the consistency of their messages and marketing which, in the case of Jaguar, has not been consistent over the thirty years of their motorsport activity. Whether Jaguar will now pursue a consistent line and continue with the essential characteristics of Clarity, Consistency and Leadership, remains to be seen.

Implementation of the Campaigns

Subaru: From the outset of the Prodrive and Subaru partnership the two organizations had a relatively well co-ordinated set of criteria for measuring success, and most importantly Prodrive delivered success in rallying and the WRC very early on and maintained a level of success that provided a marketing platform (see Chapter 4). FHI and Subaru, early in their campaign, developed a worldwide set of promotional activities that included their distributors and dealers, as well as the marketing and sales departments of the company in Tokyo and around the world (Chapter 5 p. 72). The main areas of activity were as follows:

1. Worldwide promotional activities co-ordinating with national dealerships.
2. Advertising and brochure campaigns emphasizing the participation and success of the team in the WRC.
3. Media coverage in all areas (press, TV and so on) involving the WRC campaign.
4. Specific dealership promotions using the rally cars.
5. Promotional activities at WRC and Group N Rally events involving dealers and distributors.

The very nature of the campaign and the consistency of it, gave FHI and

Subaru the opportunity to continually develop their marketing efforts and made it possible for the company to fulfil long term objectives. Rally success gave the company a steady stream of new material, and underlines the fact that continued participation is not enough without competition success.

In the case of FHI and Subaru the level and consistency of management over a period of more than a decade have been remarkable. Over that length of time the company has sustained a very clear delineation of responsibilities between STI and Prodrive. Both the philosophy and the actual activities on the programme of development and rallying have been continually agreed and recognized by both parties.

Jaguar: Throughout their thirty years of motorsport campaigns Jaguar have had tremendous success, intermittently. It is true that their focus has always been on only entering competition if they are sure to reap a VFM, value for money, result (Chapter 10, p. 126). In the event they usually have, although the latest campaign in Formula 1 is costing the company more than any previous campaign (See Chapter 11, p. 150) with, as yet, disappointing results and Formula 1 is almost certainly a long term exercise.

In one other context the two companies appear to be approaching the current activities rather differently, that of managing and controlling their campaign efforts. FHI and Subaru (see Chapter 3) sustained over more than a decade a very clear delineation of responsibilities between STI and Prodrive.

With regard to their current Formula 1 programme, in a short space of time, Ford and Jaguar have had to face a number of managerial changes, some of which have been forced upon them and some which have been taken voluntarily. But the net impact seems to be a certain level of uncertainty and lack of strategic direction, making it more difficult for the Jaguar team to deliver the necessary success. Furthermore, the company has been rightly criticized for the lack of results on the track, particularly in view of the performance of the Stewart Team in the season before Ford took over (Chapter 11, p. 147). In spite of the costs of participating, the message from Jaguar and Subaru is very clear, that simply participating is not enough.

Marketing Results

Subaru: In the company's 1990 Annual Report, the President and Chairman of the company Toshihiro Tajima reported that 'the sales of the Legacy had risen from 46 724 in 1988 (the year if its launch) to 63 793 by 1991.' In addition, he also reported that 'the Legacy series continued to expand its reputation in world markets' and that 'Sales of the Automotive Division increased 16.7 per cent, thereby generating '81.0 per cent of the Company's net sales'. This was in spite of the continued recession in the United States and Canada and, although there was an overall decline in total passenger car sales in the European market, Subaru's unit sales rose by 13.7 per cent.

This was the first year of the company's involvement in the World Rally Championship and the Legacy Model had already made its impact on the motorsport world. The Subaru World Rally Team had taken places in the Acropolis Rally in Greece, the Safari Rally, the Thousand Lakes Rally in Finland and rallies in the United States, Australia and New Zealand. These activities had helped to bring the new car to the attention of a wider motoring public and achieved greater market exposure than would have been possible by other means. In these early days the annual budget was in the region of £9 m and was funded almost entirely by Fuji Heavy Industries.

The company was able to progressively increase the sales of its Legacy model against the trend in both the domestic and overseas markets. In particular the company was replacing sales of the low value added mini vehicles with the premium priced larger vehicles such as the Legacy (1989–94) and the Impreza from 1992.

The company's rally successes and related marketing efforts, at least from raw sales statistics, appear to have had an impressive impact upon both domestic and overseas sales. The company's sales revenues moved from being reliant on low value added mini vehicles and their domestic market to a sales profile dominated by higher value added vehicles and overseas sales. In 1992, sales of the company's main models, the Legacy, Impreza and Forester, rose from a total of 83 584 (domestic) and 239 929 (overseas) units, which was 58 per cent of total sales. By 1999 these figures had increased to 124 341 (domestic) and 271 218 (overseas) units, which was by then 69 per cent of total sales. There is little doubt that some of the increase in overseas sales was due to the decline of the Yen between 1997 and 1998. However, the trend has continued into 1999 in spite of the Yen's rise against the $US; from July 1999 to July 2000 the Yen rose from ¥122/$ to ¥107/$ (see Chapter 5).

Jaguar: In the early years of Jaguar's marketing campaigns, motorsport was a key element in establishing the marque's pedigree and performance credentials. Under the management of William Lyons, motorsport was never an indulgence and was undertaken with the clear objective that the outcome had to be seen in the showroom as well as on the track; without the former, competition was a waste of resources. Nevertheless, the sales results are impressive and from the early beginnings in the immediate post-war period the five Le Mans victories from the first in 1951 certainly helped to push the sales figures over 10 000 and then over 21 000 by 1960. Whilst there is little doubt that the booming export markets of the period were a major factor in the company's success, the marketing efforts that were built around the company's competition programme were financially effective (see Chapter 7).

The company was never in motorsport for any other reason than to boost the sales figures and by 1956 the time to pull out was seemingly appropriate (Chapter 7, p. 97). The accident at Le Mans in 1955 was a major factor in their withdrawal from racing, but there were other factors involved, including the decision by Mercedes to cease their racing programme. The company did not succeed in improving sales through competition again until two decades had passed. The abortive attempt to return to saloon car racing in the days of British Leyland's ownership did nothing to enhance the brand's image and we could discern no evidence of any sales improvement at all (Chapter 7, p. 98). That particular campaign is extreme evidence that mere participation is definitely not enough to add advantage, indeed failure can damage the brand. Jaguar then withdrew from official support of any competition. The next phase of the Jaguar motorsport marketing enterprise started in the early 1980s as the company struggled for survival, and with the fact that it was essentially a one-product company. In May 1980 John Egan became chairman of Jaguar under Leyland rule and he recalled: 'Do you know we were just a one product company then? For my first nine months we made no XJ-S coupes at all!' The Canadian importers ended that drought with an order for 100, Egan remembered, 'then we made another 100 and that seemed to saturate the market for a while'.

By 1982 Egan was convinced that in order to rescue the brand, and probably the company, a new marketing initiative was required to breathe new life into the brand and to help revitalise the product. In an extraordinary deal Egan entered a 'no win no pay' arrangement with Tom Walkinshaw in 1982 to contest the European Championship. The partnership produced success at an early stage, and in the first year of the partnership between TWR and Jaguar, the XJ-S cars won four races and came second in the Championship. In the following year, 1983, Jaguar officially supported the effort for the first time, and the cars again came second in the Championship.

In the following years decade the partnership of the two companies presented Jaguar with notable successes and opportunities for extremely effective marketing campaigns. It is estimated the campaign cost Jaguar some £300 000 per annum from 1983 to 1984, although the sponsorship amounts provided the rest of the estimated £5 m annual budget needed to run the World Championship winning XJR-8 and 9 Southgate-designed cars in 1986–91. The contribution of that expenditure to the revival of the brand and the continuing sales of the XJ-S model is obvious (see Chapters 8 and 9).

The phenomenal success of the campaigns is evident in the sales figures of the 'doomed XJ-S' in 1980 and afterwards. The product's sales in 1980 were only 1760 vehicles, but after more than a decade of production more than 10 000 vehicles were sold in 1988 and when production ceased in 1996 a total of 115 413 units had been sold (see Chapter 10). Although there were difficulties in the later stages of the relationship, particularly concerning the production of road cars, the motorsport campaign of the 1980s and early 1990s made a substantial contribution to the success of the marque. Finally, there is no doubt that the campaign made a significant difference to the share price that Ford Motor Company eventually paid for the privatized company in 1989 (See Chapter 10).

There are no discernible trends in sales from the results of the current Jaguar Formula 1 campaign and it is too early to make any comparisons between the two current programmes. We have set out the manner in which the company's first motorsport activity since the end of the TWR partnership was established (Chapter 11). However, two aspects of the current Jaguar Formula 1 efforts do appear to be important. Firstly, in motorsport money does not guarantee success and in fact never has done, a fact not confined to motorsport. It does appear that Ford have been somewhat active in applying too much management to the situation. Secondly, the Stewart Grand Prix team had made an impact on the front rows of the grid and achieved remarkable results in their first two seasons, results which Ford and Jaguar have failed to build upon, in spite of an increased budget.

In terms of management style and implementation, it is still to be proved that a large corporation can effectively run a racing team. It is no surprise to the authors that the transition from personal ownership and management to corporate ownership has been difficult (See *Britain's Winning Formula*, pp. 258–9). The large number of management changes over a short period of time are one probable reason why the delicate cohesion of the team may have been disturbed. On such vitally important factors as the effectiveness of the team, which are difficult to define and generate, may the actual performance on the track depend: particularly so for a team which has access to all the necessary technical resources.

Conclusions: Branding Lessons for the Future

The work of sports sponsorship in elevating brand value and, of course, profits is a factor in modern corporate marketing that has been in the marketeers' armoury for several decades. In common with many marketing activities it is exceptionally difficult to isolate the impact of expenditure, and therefore the benefits derived. We have detailed the activities and, more importantly, the budgets allocated by two corporations attempting to achieve success in one of the most competitive markets in the world. Over-capacity, stable or at the most very slow growth in the world's major markets, capacity is still increasing in the first year of the new millennium. Technology is increasingly diffused, and the ability of manufacturers to sustain a competitive advantage through technology more difficult. Differentiating products with similar performance and features, is an increasing pre-occupation of the marketing effort, in order to generate the opportunity for premium prices, customer satisfaction and loyalty.

Competitive motorsports are an opportunity for the appropriate manufacturer to 'engage' existing and potential customers in the products and to increase the degree of affinity. In addition there is little doubt, as the evidence contained in this book supports, of the efficiency and effectiveness of motorsports as a marketing tool. It is efficient in the degree of coverage and access to consumers that it provides and in the manner in which it exploits the supplier's relationship with the manufacturer to generate partnerships to provide the funds to market the brand.

These two campaigns were undertaken in extremely different circumstances and with different plans. The problem for Subaru was that they had little or no market presence and certainly no 'brand differentiation' in the markets they had to conquer. But the problem is that niche market products can, in some circumstances, be replicated, as advances in product and design technology speeds the process of product innovations. Whilst it is true that the keys to success for the low volume manufacturer, are to move from high volume, low value added models to products that:

- Improve the value equation
- Separate form from function
- Achieve joy of use

Gary Hamel

Even relatively large volume players are having to take the same road to success as the market place becomes ever more crowded with similar products. The rationale behind the Aston Martin DB7 was precisely to be able to sell more vehicles to a larger and younger market, and to increase the brand's profile worldwide. The problem of differentiation will always be there and both companies have recognized that fact. As Mr Takemasa Yamada expressed it: 'Giant makers soon imitate and we lose originality and competition'.

Achieving differentiation is a constant battle and one that demands a consistent and attentive approach to the brand and the qualities of the products; the essence of the brand. There are some important findings that have been established from an analysis of visions of the future of brands from a number of brand experts (see. *The Future of Brands* by Interbrand (Macmillan, ISBN 0–333–77673–9). Some of the most important of these have been confirmed by our study.

A Brand with No Clear Vision has No Clear Future

In the early discussions at FHI on their marketing problem it was apparent that the organization was very clear about their vision of the brand they wanted to create. Furthermore, they were very focused on the characteristics of the products and the consumers who would purchase them: 'instinctively aggressive and competitive and wanting a product that appeals to their instinct'.

'The cars offer SPEED, STYLE, SPACE'.

Source STI

Of course it may have been fortuitous and beneficial that Prodrive was, and is, also a company with a distinct, consistent and focused vision.

In their marketing and competition activities, Jaguar have always focused upon the characteristics of the product and the values that the brand identified. It is in the more recent developments under the Ford banner that the relationship between the vision and the reality is somewhat distorted. Participation in competition signifies that the performance has to mirror the product vision, it is an inadequate base on its own.

Values-led Marketing Will Create Stronger Brand Relationships

Both companies have had a clear set of values although the surprising fact in the FHI and Subaru story is the relentless consistency of the message derived from both the marketing and the product.

Jaguar has been less consistent in terms of product quality and the marketing message.

However, the other part of this message is whether the values of a brand can remain unchanged or do they have to reflect the changing values of society and the consumer? In particular, the values expressed by the Subaru brand are aggressive and open to interpretation in a manner that is distinctly outside the frame of values held by many consumers.

It is Increasingly Important to Understand What Makes a Brand

Valuable

The financial value of the brand development that has taken place over the time-scale in which both case studies have been considered is clear from the analysis that we have undertaken. In the case of the Jaguar brand that value has, at least in part, been the subject of the judgement of shareholders and therefore important stakeholders. In our discussions with FHI and from all the sales and market evidence, there was certainly no doubt that the company regarded the value of the brand as a direct function of their investment in motorsport.

However, in both cases the brand's value is far more than the sum of the added sales or the increased share price, at a specific time. Brands have a significant time value in that they are an asset in their own right and therefore they require managing in the same way as any other asset; to maximize its value to the organization. In the case of Ford and Jaguar, it will be interesting to see whether the underlying corporate values of both corporations are mutually sustainable. The acquisition of Aston Martin and Jaguar gives the company a brand management issue, and makes understanding of the value the brands bring to the corporate whole a crucial factor in enhancing shareholder value. Managing different companies with different brands means managing different, and in Ford's case diverse, sets of values.

There are three other issues that we believe are illustrated by the experience of the two cases. These are the vital importance of the following factors when pursuing any programme or campaign that is meant to add value to a brand.

Clarity – of vision, mission and values, which are understood lived and even loved by the people who deliver them. Clarity of what makes those values distinctive and relevant; and clarity of their ownership in both people's minds and trademark law around the world.

Consistency – not in the sense of a simple product, nor in the sense of predictability. Successful brands are consistent in the values, concepts and level of quality they or their products deliver to the consumer. A consistent alignment of values such as that expressed by Moet & Chandon, Coca Cola, Ford, BMW or Gillette to name but a few.

Leadership – a consistent characteristic of successful brands is the brand's ability to lead and exceed expectations, to take people into new territories and new areas of product, service and even social philosophy at the right time. It is about a brand's ability to be restless about self-renewal.

(From *The World's Greatest Brands*)

Finally, in both the cases we have described the fact remains that:

The brand sets the pace but the product has to be able to fulfil and even extend the brand's image and attributes, benefits and essence: the values of the brand.

Index